LEARNING MATHEMATICS:
CONSTRUCTIVIST AND INTERACTIONIST THEORIES OF
MATHEMATICAL DEVELOPMENT

LEARNING MATHEMATICS

*Constructivist and Interactionist Theories of
Mathematical Development*

edited by

PAUL COBB

*Peabody College, Vanderbilt University,
Nashville, Tennessee, U.S.A.*

Reprinted from Educational Studies in Mathematics 26 (2–3), 1994

KLUWER ACADEMIC PUBLISHERS
DORDRECHT / BOSTON / LONDON

Library of Congress Cataloging-in-Publication Data

A C.I.P. Catalogue record for this book is available from the Library of Congress

ISBN 0-7923-2823-X

Published by Kluwer Academic Publishers,
P.O. Box 17, 3300 AA Dordrecht, The Netherlands.

Kluwer Academic Publishers incorporates
the publishing programmes of
D. Reidel, Martinus Nijhoff, Dr W. Junk and MTP Press.

Sold and distributed in the U.S.A. and Canada
by Kluwer Academic Publishers,
101 Philip Drive, Norwell, MA 02061, U.S.A.

In all other countries, sold and distributed
by Kluwer Academic Publishers Group,
P.O. Box 322, 3300 AH Dordrecht, The Netherlands.

TABLE OF CONTENTS

PREFACE

The first five contributions to this Special Issue on Theories of Mathematical Learning take a cognitive perspective whereas the sixth, that by Voigt, takes an interactionist perspective. The common theme that links the six articles is the focus on students' inferred experiences as the starting point in the theory-building process. This emphasis on the meanings that objects and events have for students within their experiential realities can be contrasted with approaches in which the goal is to specify cognitive behaviors that yield an input-output match with observed behavior. It is important to note that the term 'experience' as it is used in these articles is not restricted to physical or sensory-motor experience. A perusal of the first five articles indicates that it includes reflective experiences that involve reviewing prior activity and anticipating the results of potential activity. In addition, by emphasizing interaction and communication, Voigt's contribution reminds us that personal experiences do not arise in a vacuum but instead have a social aspect.

In taking a cognitive perspective, the first five contributions analyze the processes by which students conceptually reorganize their experiential realities and thus construct increasingly sophisticated mathematical ways of knowing. The conceptual constructions addressed by these theorists, ranging as they do from fractions to the Fundamental Theorem of Calculus, indicate that experiential approaches to mathematical cognition are viable at all levels of mathematical development. Although the authors use different theoretical constructs, several additional commonalities can be discerned in their work. For example, all would seem to concur with Steffe and Wiegel's claim that the learning environments that students establish are a function of the concepts and operations they use when interpreting situations. Further, all are concerned with what is colloquially called meaningful learning – learning that involves the construction of experientially-real mathematical objects. In addition, all trace the origin of these mathematical objects to students' activity, both physical and conceptual. Thus, Sfard and Linchevski discuss the process of reification, and Steffe and Wiegel analyze students' internalization and interiorization of their activity. In taking this approach, the authors characterize knowledge as action – they are concerned with students' mathematical ways of knowing. Pirie and Kieren, for example, stress that mathematical understanding is a process, not an acquisition or a location. Similarly, Confrey and Smith differentiate between the various types of additive and multiplicative units that students construct in terms of the actions that generate them. Thompson,

Educational Studies in Mathematics **26**: 105–109, 1994.
© 1994 *Kluwer Academic Publishers. Printed in the Netherlands.*

for his part, stresses that his focus is on the conceptual operations and imagery that constitute students' understanding of particular mathematical topics.

A further commonality that cuts across the contributions concerns the importance attributed to imagery. For example, Steffe and Wiegel talk of students representing their mathematical activity by running through it in thought. Confrey and Smith refer to an experiential dimension wherein children distinguish more, less, and the same rates of change non-numerically, perhaps when sensing a change of speed while riding in an automobile. In Confrey and Smith's view, imagined sensations of this type constitute one of the bases from which more sophisticated conceptions of rates of change develop. Consistent with the general emphasis on knowing rather than knowledge, both Thompson and Pirie and Kieren argue that images are dynamic aspects of the understanding process rather than static mental pictures. Both also stress that the role imagery plays in mathematical activity evolves as particular concepts become increasingly abstract. Thus, Thompson draws on Piaget's work to distinguish between three types of images. The third of these types of images appears to be reflexively related to the student's mental operations in that the image is shaped by the operations, and the operations are constrained by the image. Similarly, in discussing a process termed folding back that occurs when the student is faced with a problem or question that is not immediately solvable, Pirie and Kieren argue that the resulting image is not the same as images formed earlier in development because it is informed by more abstract interests and understandings. It should be noted that the general focus on imagery is not restricted to elementary mathematical concepts. Thompson in fact locates students' difficulties with the Fundamental Theorem of Calculus in their impoverished images of rate, and Pirie and Kieren argue that more advanced mathematics needs to be worked out at the level of image making before students can begin to look for formalization or structure.

An additional point of contact between the authors concerns the pragmatic approach they take to theorizing. Thus, Thompson argues that a theory is an anticipatory scheme that enables the researcher to imagine problems that might arise in the learning-teaching situation, and to plan potential solutions to them. In a similar vein, Steffe and Wiegel say that their overall objective is to develop what they call the technical knowledge necessary for teaching. This theoretical pragmatism is reflected in the methodologies used by the researchers. All involve the intensive analysis of students' mathematical activity, resulting in theoretical constructs that are empirically grounded in a non-empiricist sense.

Both Thompson's and Sfard and Linchevski's contributions indicate that historical analyses complement experiential approaches to cognition. Thompson's motivation for analyzing Newton's and Leibniz's work is to look for kinds of reasoning that might provide a starting point for instruction oriented to the development of imagery and forms of expression that could support later insights into central ideas of the calculus. Sfard and Linchevski's objective in analyzing the historical development of algebra is to identify possible developmental stages that both clarify students' difficulties and inform instruction. In light of this demonstrated relevance of historical analyses, Sfard and Linchevski are careful to note that caution should be exercised when drawing analogies between historical and psychological development because a deliberately guided process of reconstruction might not necessarily follow the meandering path of the first travelers.

A further emerging theme apparent in the contributions concerns the role that language and symbols play in mathematical development. Sfard and Linchevski's historical analysis of algebra indicates that the introduction of symbols is a central aspect of the reification process by which mathematical activity is objectified. While acknowledging that the development of symbols is by itself insufficient, they nonetheless contend that symbols are manipulable in a way that words are not, and this makes it possible for algebraic concepts to have an object-like quality. Pirie and Kieren make a related argument when they claim that both acting and expressing are necessary at any level of understanding before a student can move on to another level. Their observations of students indicate that in the absence of an expression of understanding, mathematical notions based solely on acting are ephemeral and do not remain with students from one session to the next. This leads Pirie and Kieren to conclude that acting and expressing constitute a complementarity at each of the levels of understanding they have identified. Confrey and Smith's analysis of the various types of units that students construct while solving problems that involve rates of change succintly illustrates this complementarity. They note that unit names from the additive counting world are not relevant to a multiplicative splitting world such as that in which a student shares a pie by splitting it in half, and then splits the halves in half, and so on. As part of their analysis, they develop names for the process of generating units in this multiplicative world. For example, they describe the process by which a student cuts a pie into eight equal pieces as three multiplicative units of two, or three 2-splits.

This discussion of symbolizing and expressing hints at the social aspect of mathematical development. Thus, Pirie and Kieren note that students express their understanding for others as well as themselves. More generally, ways of symbolizing and expressing are negotiated in the course of social interactions. It is in regard that Voigt's presentation of his interactionist perspective complements the cognitive perspective taken in the first five papers. Voigt's basic claim is that the social dimension is intrinsic to mathematical development because the teaching-learning process is an interpersonal process. Both Steffe and Wiegel, and Pirie and Kieren touch on this social dimension when they discuss the influence of the teacher's interventions on students' goals and interpretations. Voigt views these interventions as actions in an ongoing process of negotiation and argues that mathematical meanings are necessarily matters for negotiation because every object and event is potentially ambiguous and plurisemantic. Voigt also contends that negotiation occurs even when the teacher and students do not explicitly argue from different points of view. Further, the sample episodes he presents indicate that it is not only the students who develop novel understandings in the course of these interactions, but the teacher also modifies his or her interpretations while negotiating meanings. This point is made particularly forcefully by Confrey and

Smith when they note that they, as teachers, reformulated their own mathematical understanding as they interpreted students' approaches and methods in the course of a teaching experiment.

Several of Voigt's examples deal with the process of mathematization by which an empirical situation is transformed into a mathematical one. Steffe and Wiegel also address this issue in some detail when they discuss how a student's cognitive play is transformed into mathematical activity. A comparison of these analyses highlights the complementary nature of the cognitive constructivist and interactionist perspectives. Steffe and Wiegel's focus is on the ways in which the researcher influences the student's interpretations during a teaching experiment. Their analysis is made from the perspective of someone who is inside the interaction and is concerned with the ways in which the student modifies his or her own activity in the course of the interaction. In contrast, Voigt's analysis is made from the outside, from the point of view of someone who is an observer of rather than a participant in ongoing interactions. From this perspective, the focus is on the taken-to-be-shared meanings that emerge between the teacher and students rather than on the meanings of any individual participant. These taken-to-be-shared meanings can be thought of as constituting an evolving consensual domain for mathematical communication. As Voigt makes clear, taken-to-be-shared meanings are not cognitive elements that capture, say, the partial match of individual interpretations but are instead located at the level of interaction. Thus, whereas for Steffe and Wiegel, mathematization is a process of individual conceptual reorganization, for Voigt it is a process of negotiation in the course of which the teacher and students collectively modify what is taken-to-be-shared between them. The complementarity between the two positions becomes apparent when it is noted that taken-to-be-shared or consensual meanings are established by the teacher and students as they attempt to coordinate their individual activities. Conversely, the teacher's and students' participation in the establishment of taken-to-be-shared mathematical meanings both supports and constrains their individual interpretations.

Both Sfard and Linchevski's and Thompson's contributions draw attention to a further feature of the social aspect of mathematical development. Sfard and Linchevski report that the students in their study frequently evidenced semantically-debased conceptions in which algebraic formulae were viewed as nothing more than strings of symbols to which certain procedures are routinely applied. Similarly, Thompson observed that the supposedly mathematically-sophisticated students who participated in this teaching experiment tended to use notation opaquely by simply associating patterns of routine figural actions with various notational configurations. Thus, as Thompson notes, although his primary focus was on students' evolving cognitions, he found himself negotiating what it means to know and do mathematics with them. In this regard, Voigt stresses that students' learning is not restricted to mathematics. They also learn how to negotiate mathematical meanings with the teacher and, in the process, develop beliefs about what counts as a problem, a solution, an explanation, and a justification. Consequently, from Voigt's interactionist perspective, Thompson can be seen to be initiating and guiding the renegotiation of the norms that constitute the classroom mathematical microculture. This process of renegotiation appears to have been crucial given Thompson's interest in meaningful mathematical activity that involves attempts to interpret notation by establishing imagery. Thompson's experiences indicate both the value of long-term teaching experiments and the need to consider the social dimensions of mathematical development even when the research focus is primarily cognitive.

One final issue that cuts across all six articles is that of attending to what Confrey and Smith call the student's voice. Relatedly, Steffe and Wiegel argue that students will not sustain mathematical activity unless they experience satisfaction in the course of that activity. The transcripts presented in the five cognitive papers indicate that, for the most part, the students did experience satisfaction as they engaged in mathematical activity with the researchers. Voigt, speaking from the interactionist perspective, observes that students are necessarily active participants who make original contributions to the establishment of consensual meanings. As the cognitive papers indicate, instructional approaches compatible with constructivism attempt to bring this originality to the fore by viewing it as a resource to be capitalized on rather than as an impediment to be pushed underground.

Peabody College   PAUL COBB
Vanderbilt University
Nashville, TN 37203
U.S.A.

LESLIE P. STEFFE AND HEIDE G. WIEGEL

COGNITIVE PLAY AND MATHEMATICAL LEARNING IN COMPUTER MICROWORLDS

ABSTRACT. Based on the constructivist principle of active learning, we focus on children's transformation of their cognitive play activity into what we regard as independent mathematical activity. We analyze how, in the process of this transformation, children modify their cognitive play activities. For such a modification to occur, we argue that the cognitive play activity has to involve operations of intelligence which yield situations of mathematical schemes.

We present two distinctly different cases. If the first case, the medium of the cognitive play activity was a discrete computer microworld. We illustrate how two children transformed the playful activity of making pluralities into situations of their counting schemes. In the second case, the medium was a continuous microworld. We illustrate two children's transformation of the playful activity of making line segments ("sticks") into situations of their counting schemes. We explain one child's transformation as a generalizing assimilation because it was immediate and powerful. The transformation made by the other child was more protracted, and social interaction played a prominent role. We specify several types of accommodations induced by this social interaction, accommodations we see as critical for understanding active mathematics learning. Finally, we illustrate how a playful orientation of independent mathematical activity can be inherited from cognitive play.

INTRODUCTION

Is there a model of mathematical learning powerful enough to be useful to mathematics teachers in their teaching? Is it enough to agree that "knowledge is not passively received but actively built up by the cognizing subject" (von Glasersfeld, 1989, p. 162)? According to Thompson (1991),

constructivism ... is seen by many as being more useful as an orienting framework than as an explanatory framework when investigating questions of learning (p. 261).

von Glasersfeld's principle of active learning is orienting, but it was not intended to provide the technical knowledge about mathematics learning critical in teaching mathematics. Our goal in this paper is to portray what such technical knowledge might look like in an evolving constructivist model of mathematical learning. Children's play is one feature of this model because children construct much of their reality through playing. More specifically, playing in a mathematical context could serve in

Educational Studies in Mathematics **26**: 111–134, 1994.
© 1994 *Kluwer Academic Publishers. Printed in the Netherlands.*

children's construction of a mathematical reality and as a source of their motivation to do mathematics.

Motivation, a key element in any model of learning (Hilgard and Bower, 1966), is often dichotomized into intrinsic and extrinsic motivation. In constructivism, the study of mathematics is regarded as intrinsically motivating. Although external reinforcement is not completely rejected, it cannot take the place of the internal satisfaction achieved from understanding mathematics.

One thing that is often by far the most reinforcing for a cognitive organism [is] to achieve a satisfactory organization, a viable way of dealing with some sector of experience (von Glasersfeld, 1983, p. 65).

We question, however, whether the internal satisfaction gained from understanding mathematics is sufficient for children to want to engage in mathematical activity and to independently sustain it over a long period of time. Consider, for example, Johanna, a third-grade student (Steffe and von Glasersfeld, 1985), who used strategic reasoning to organize adding and subtracting whole numbers. Asked which number should be added to 12 to obtain 19, she first added 9 to 10 to obtain 19 and then subtracted 2 from 9. Strategic reasoning to find sums and differences is certainly a viable way of organizing adding and subtracting, and the relationships it produced seemed indeed to be satisfying to Johanna. But her reasoning had not evolved from her play activities, nor was her orientation one of play. Instead, her strategic reasoning had been induced by the teacher's question in an obligatory rather than in a playful context, and as such, her strategic reasoning was compulsory. Had her reasoning been playful, the satisfaction she experienced might have been intensified, and her numerical strategy might have become a part of her construction of a mathematical reality. In that case, Johanna might have viewed strategic reasoning as belonging to her, and it would have been more likely for her to repeat it in a variety of contexts for its functional pleasure.

von Glasersfeld's proposition does not exclude a playful orientation of a cognitive activity – such as strategic reasoning – if it is ready-at-hand (an activity defined by assimilation), and if it is carried out "purely for functional pleasure (Piaget, 1962, p. 89)." How to foster such a playful orientation of children's mathematical activity is part of the technical knowledge about mathematics learning we believe should be common among mathematics teachers. Fostering mathematical activity with a playful orientation can be challenging if children – like Johanna – regard their school mathematical activity as compulsory. In our teaching, we first engage children in cognitive play activities with the goal that the children establish the enactment of basic conceptual operations as pleasurable. We then intervene in an attempt to transform the cognitive play activity into a mathematical activity while trying to preserve the children's playful orientation.

We illustrate such a cognitive play activity and what we have found to be essential elements in its transformation into mathematical activity. The elements we discuss are: (a) the nature of the medium in which children play and the basic conceptual operations they enact in the medium, (b) the schemes of action and operation children use in the transformation, (c) the involved accommodations, and (d) the role of social interaction in inducing these accommodations. These elements are all involved in establishing technical knowledge of mathematical learning.

MICROWORLDS: A DYNAMIC MEDIUM

The nature of the medium in which children learn mathematics is essential because the medium must not only support the construction of mathematical concepts and operations, it must also provide the possibility for creative expressions of those concepts and operations in the pursuit of goals. We find Papert's (1980) idea of a microworld a promising approach for creating a dynamic medium in which children can learn mathematics and creatively express what they have learned. A microworld is a self-contained world in which children "learn to transfer habits of exploration from their personal lives to the formal domain of scientific construction" (p. 117). Using this general idea of a microworld within our project, we created several computer microworlds in which children might engage in cognitive play activity and might learn to play mathematically.

Our microworlds consist of objects children can make and of possible actions they can use to operate on the objects. In the design of the microworlds, we built the actions to be compatible with our understanding of children's numerical and quantitative operations (Steffe et al., 1983; Steffe and Cobb, 1988; Steffe, 1992). We do not claim, however, that the possible actions are identical to children's numerical operations nor do we claim that by performing the actions, children necessarily perform their numerical operations. Quite on the contrary, we designed the actions of the microworlds so children could perform them without performing any numerical operations. Nevertheless, executing the possible actions of the microworlds could be instantiations of the numerical operations children use in pursuing their numerical goals.

Cognitive Play in Toys

In the microworld *Toys*, a child can place a toy in the playground by choosing any of five small geometrical figures (triangle, square, pentagon, hexagon, heptagon) from the toy box and then clicking the mouse pointer inside the playground. After choosing a toy, the child can make replicates of it by repeatedly clicking in the playground. This basic operation of replicating toys introduced a repeatable action our students turned into activities fitting Piaget's (1962, p. 89) definition of play as an activity repeated for functional pleasure. Working in pairs, the children converted the activity of making and replicating toys into creative ventures. Some of them made what appeared to be space curves by overlapping the toys and crossing the evolving paths back and forth. Other children drew faces, cars, bicycles, jack-o-lanterns, and other recognizable figures. Still other children made stacks of toys. Although the activity of repeatedly clicking was itself pleasurable, the extension and organization of that activity into the creation of designs contributed notably to that experience of pleasure.

There were several aspects of the children's actions fitting Piaget's (1962, pp. 147–150) criteria of play. First, we as the children's teachers made no suggestions of what the children should do beyond an initial demonstration of the basic computer commands. Whatever the students made was without directives and appeared to be spontaneous. Second, the spontaneous activities were pleasing to the children. Any partially completed design or drawing seemed to lead to additional possibilities. The designs became visually appealing, as they were being made and in their completed forms. Making replicate toys seemed to stimulate the students' imagination, and they often created a design or drawing not foreseen at the beginning of the activity. Third, there were no constraints on what the children should or should not make; they were free of conflicts. Also, the production of toys was an organized activity. The children established order in the replication of toys and seemed to take great satisfaction in doing so. The play activity allowed the children to experience the dynamic quality of Toys and to establish the use of the microworld as a pleasurable activity.

The act of replicating a toy was an enactment of those mental operations that produce a sequence of sensory items, a sensory plurality. The enactment of mental operations was one reason why this activity – the production of sensory pluralities – was a pleasurable activity. Another reason was that the production involved the enactment of mental operations yielding more sophisticated composite wholes (von Glasersfeld, 1981). The enactment of these mental operations provides a specific meaning for Piaget's (1962) comment that almost all of the behaviors we studied in relation to intelligence are susceptible of becoming play as soon as they are repeated for mere assimilation (p. 89).

If we understand the production of sensory pluralities as a fundamental operation of intelligence on which the construction of concepts (including numerical concepts), composite units, number sequences, and more general quantitative reasoning (Steffe, 1991a) is based, we can more fully appreciate the children's free play activity as enactment of these abstracted structures.

Activating Counting in Cognitive Play

The students' creation of pluralities in the Toys microworld provided us with a starting point to transform children's cognitive play activity into a mathematical play activity. If a cognitive play activity is to be transformed into mathematical activity, this play activity has to involve operations of intelligence yielding a situation of a mathematical scheme. We demonstrate the transformation with an example of Melissa and Courtney. In an initial intervention, the teacher asked Melissa to cooperate with her partner and to stop him when he had made no more than 50 toys. Courtney clicked the mouse swiftly so Melissa could not keep an accurate count. This fast production of toys was intentional on Courtney's part, demonstrating he was still playing. When Melissa stopped him, Courtney had made 58 toys. Melissa, asked to make it right, dragged the extra eight toys into the trash.

Reversing roles, the teacher asked Courtney to guess how many toys Melissa made. When Melissa stopped putting toys into the playground, Courtney guessed 69 toys. The children asked the computer how many toys were in the playground, and the computer's number box showed 58 toys. The teacher asked:

Protocol 1

Teacher: How many more would you need to make sixty-nine?

Courtney: (Tries to continue making toys, but is stopped by the teacher. Then) two more!

Teacher: Do you have a way to find out?

Melissa: I know. (Sequentially puts up fingers) fifty-nine, sixty, ... sixty-nine. Eleven.

Melissa's break at 58 was an opportunity to ask a question which might activate the children's counting schemes. Had Courtney not been stopped from using the mouse to produce more toys, he may have continued the activity until he had reached 69. This would have constituted an activation of his counting scheme, but not in the same way Melissa's counting scheme was activated. Whereas Courtney would have continued to work on a perceptual level, Melissa reconstituted the situation as a numerical one.

Using her numerical operations, Melissa reconstituted the plurality established when activating the computer commands, as a mathematical situation. As she interacted with the teacher and responded to his constraint, her situation changed to include her number sequence. From our interview with Melissa at the beginning of the teaching experiment, we already knew she had constructed a number sequence, and it had been our goal for her to become aware that the number sequence was relevant in the context of Toys. Now that Melissa had established this relevance in the context of a play activity, her counting activity had a playful orientation. We repeatedly observed such a playful orientation during the first year of the teaching experiment. A playful orientation of children's mathematical schemes of action and operation is especially important if one regards the children's use and modification of the schemes as the mathematics of education.

Melissa's transformation of a plurality of toys into a situation of her number sequence illustrates that the learning environments children establish are a function of the concepts and operations they use in establishing them. This principle is fundamental in a constructivist model of learning; it defines as one task of the teacher to encourage children to become aware of their mathematical schemes and of the situations in which they can be used.

ACCOMMODATIONS IN CHILDREN'S NUMBER SEQUENCES

Some of the most important modifications of children's number sequences occur when children make the extension from discrete to continuous situations. Studying how children make this extension is one of our goals in the project. This goal is important to us because we intend to explore children's construction of fractions in continuous situations while taking advantage of their whole number knowledge. A microworld we call *Sticks* allows the creation of continuous units. In Sticks, children produce colored line segments (sticks) by activating the DRAW command and then sweeping the cursor horizontally over the screen. From the observer's perspective, this action can establish an *experiential continuous unit*.

Continuous Units

An experiential continuous unit is a perceptual unit item whose material can be formed by moving a finger through sand, a pencil on paper, a hand through space, or the eyes from one location to another. All of these movements generate experiences that, if abstracted from the rest of the experiential field by applying the unitizing operation (Steffe et al., 1983), yield experiential continuous units. The experiential continuous units children can make in Sticks are, like the pluralities in Toys, products of intelligence, and children can establish them for functional pleasure.

Figural continuous units are produced by re-presenting the experience of generating an experiential continuous unit. Being able to regenerate – without the relevant perceptual material being available in the experiential field – the trace of moving one's finger through sand, for example, is the criterion for a continuous unit to be internalized. The visual records used in regenerating this experience are an integral part of the concept of a continuous unit, but they may not exhaust the records of experience in the unit. Kinesthetic records of moving one's hand may be also involved, complementing the visual records. These records are a constitutive part of the continuous unit and may be later abstracted as the property of the unit called its length.

Interiorized continuous units are created by using figural continuous units as material of the unitizing operation. This further act of abstraction purges the continuous unit of its sensory material. But the sensory material is not gone in the sense that it is no longer contained in the unit. To the contrary, what the records in this abstracted unit pattern point to is still accessible to the individual. But the sensory material need not be activated for the individual to operate, because the figural material can now be used as material of operating.

The records of experience in an abstracted unit pattern are interiorized records as opposed to the internalized records in the figural continuous unit. Prior to this internalization, we call the records of making an experiential continuous unit perceptual to connote that the records can be instantiated only in an experiential situation. This analysis of continuous units is compatible with the distinctions among perceptual, figural, and abstract discrete units (Steffe et al., 1983).

In cases where the motion involved in establishing an experiential continuous unit is segmented, we speak of an experiential continuous but segmented unit. Likewise, we speak of internalized and interiorized continuous but segmented units. If an interiorized continuous but segmented unit becomes a situation of a child's number sequence, the child can establish the numerosity of the segments – by counting – and hence can establish a *connected number*.

We designed the computer microworld Sticks to encourage children to establish experiential continuous units, to construct internalized and interiorized continuous units, and to use their number sequences in this context. The initial activities in Sticks included drawing sticks (DRAW), changing the color of a stick (FILL), marking a stick (MARK), breaking it along the marks (BREAK), joining the parts back together (JOIN), and cutting off pieces from a stick (CUT). These activities had a playful orientation for two children in the teaching experiment, Rebecca and Tania, as they repeatedly used the commands for functional pleasure.

Re-Presentation and Imitation

There was no indication during these initial activities that the two girls would take different paths in their construction of connected numbers. However, differentiation occurred as soon as the teacher, with the intention to explore their construction of connected numbers, began to interact with the students. In retrospect, Rebecca established connected numbers almost immediately, whereas Tania took a path using re-presentation and imitation as her primary means. This difference between the students was not obvious to their teacher as he interacted with them, nor was it obvious to the witnesses of the teaching session. Only after the next teaching session did the differences become clear to us.

We illustrate Tania and Rebecca's construction of connected numbers with protocols from three successive teaching sessions. The protocols illustrate the children's accommodations in their number sequences as result of interacting with each other and with the teacher in the microworld Sticks. Because these interactions were largely teacher directed, we do not consider the children being engaged in cognitive play or in mathematical play. Rather, they were involved in situations of learning (mathematical activities) initiated by their teacher.

Given a red stick (later called the *unit stick*) about 1 inch long, the students were asked to draw a stick the same length as two red sticks. Rebecca activated COPY and made two copies of the red stick, positioning the cursor so the copies appeared end-to-end. Responding to the teacher's charge to make "one stick", she activated the JOIN command to join the copies, then erased – on her own – the mark in the middle, changed the color of the new stick to green, and moved it below the unit stick (see Figure 1).

At this point, the teacher asked how many unit sticks it would take to make the green stick.

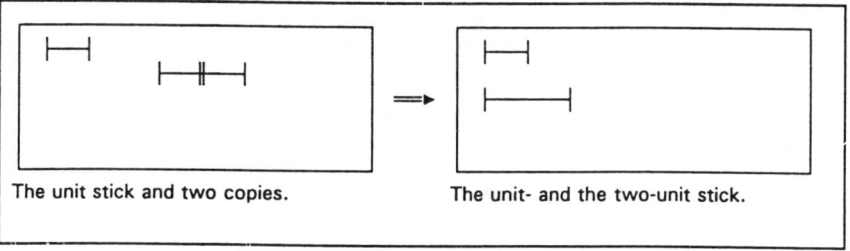

The unit stick and two copies.   The unit- and the two-unit stick.

Fig. 1.   The initial situation.

Protocol 2 (11 May 1992)

Tania: One (extends one finger).

Rebecca: (Has her hand up.)

Teacher: Rebecca?

Rebecca: Two! That's one right there (moves her finger from the
              right end mark of the unit stick down to the green stick)
              then another right there (points to the end mark of the
              green stick) and that's the same size (moves her finger
              along the second half of the green stick), so that makes
              two.

Tania: ... [inaudible], because you already had one, and then you
              put another one!

We interpret Tania's initial answer of "one" as referring to the whole
two-unit stick Tania regarded as one stick. Faced with Rebecca's conflicting solution and her explanation, Tania seemed to be justifying her own
initial response. In our inference, Tania re-presented Rebecca's actions of
making the green stick: the first copy ("you already had one"), then the
second copy ("and then you put another one"). Furthermore, she related
her answer "one" to only the second part of Rebecca's actions ("then you
put another one").

Tania's re-presentation of Rebecca's action of making and joining two
copies of the unit stick can be seen as an advanced form of imitation in
Piaget's system – a re-presentable scheme of action. Piaget (1962) defined
imitation as the primacy of accommodation over assimilation;

If the subject's schemas[1] of action are modified by the external world without his utilising
this external world ... the activity tends to become imitation (p. 5).

He linked imitation to re-presentation in that a regeneration of an immediate past experience in imagination might be described as an internalized
imitation of that experience (p.5). We consider the regeneration of an expe-

rience as imitation if there is an agent (such as another child) initiating the experienced situation and if it is possible to infer a goal on the part of the imitating child to reenact the actions of the agent. We regard imitation as a basic element without which one could not adequately understand peer interactions (Sinclair, 1990) nor interactive mathematical communication more generally (Cobb et al., 1990).

Imitation and Modification

In addition to re-presenting Rebecca's actions, Tania also incorporated her partner's explanation. In her own explanation, Tania correlated her verbal utterances with movements that were a reenactment of Rebecca's pointing acts. This incorporation led to the establishment of a re-presentable action scheme for making a figural continuous but segmented unit of two. Tania modified this scheme when she subsequently drew a stick three units long. Placing the cursor beneath the left endpoint of the two-unit stick, Tania moved it past its right endpoint without clicking off. She continued:

Protocol 3 (11 May 1992)

Tania: Right there (places her finger on the right endpoint of the unit stick and then drops it to her incomplete red stick (see Figure 2, left panel)), right there (places her finger on the right end point of the two-unit stick and drops it down to her red stick; leaves her finger on the incomplete stick directly beneath the right endpoint of the two-unit stick and continues drawing her red stick, trying to gauge where to stop, then clicks off (Figure 2, right panel)).

Teacher: Okay, can you check that?

Rebecca: (Drags the unit stick to the end of the two-unit stick, placing it end-to-end to check the length of the red stick, which is too long. She then moves the unit stick back.)

Tania, in order to complete the three-unit stick, projected the unit stick and the two-unit stick onto her incomplete stick and then visually gauged where to stop by looking back and forth between the unit stick and her new stick. Tania's actions were a coordination of the activity of drawing and of the results of the action scheme she had established through imitation. This coordination was situational in that she needed a staircase consisting of a unit stick and a two-unit stick to continuously draw a stick as long as three unit sticks. The coordinated action, occurring in the context of accomplishing a new goal, was a modification of her re-presentable scheme

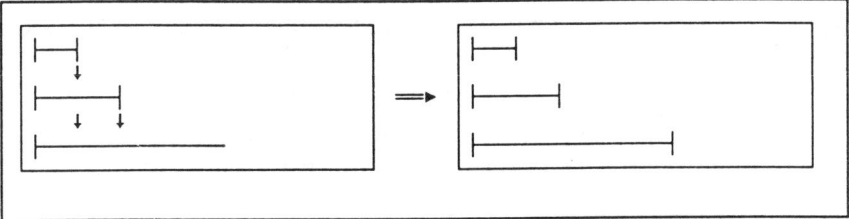

Fig. 2. Establishing a figurative continuous unit of three.

for making a continuous but segmented unit of two. This establishment of a new goal can be interpreted as "utilising this external world" (Piaget, 1962, p. 5); thus, Tania's actions were no longer imitation.

It is possible that Tania's actions of making a three-unit stick became recorded in her concept of three, establishing a connected number. However, Tania did not form the three-unit stick by independently copying the unit stick three times and then joining the copies together. Had she spontaneously done so, it would have indicated she had established a scheme for making a connected number, possibly specific to "three". When she later used the COPY command, imitating Rebecca again, she was unable to carry out all the necessary actions that would produce a six-unit stick. She made six copies of the three-unit stick – rather than of the unit stick – and joined all six copies, creating a stick that extended beyond the width of the screen. She then engaged in a series of cutting and joining actions that served to relieve her frustration rather than the goal of creating the six-unit stick. At this time, she had become very upset, and the activity had changed from being pleasurable to being stressful.

A Generalizing Assimilation

In contrast to Tania, Rebecca chose the COPY command independently. Her decision to copy the unit stick and to place the copies end-to-end (see Figure 1) indicates her concept of "two" was already activated, ready to be used in a new situation. Rebecca's explanation of why the unmarked green stick was two units long ("That's one right here, then another one right there, and that's the same size, so that makes two", Protocol 2) supports this inference. Unlike Tania, who initially referred to the two-unit stick as "one", Rebecca saw the two-unit stick as "one" (indicated by the deletion of the mark in the middle) and "two" at the same time. Having copied the two parts from the unit stick was essential for viewing them as "the same size", but it would have been insufficient had Rebecca not unitized

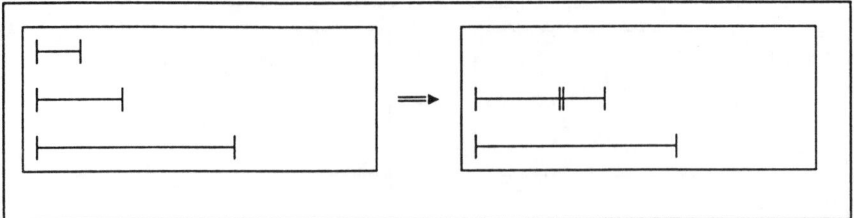

Fig. 3.   Checking the length of Tania's stick.

the two copies, creating a pair of abstract unit items containing records of the copies.

A further indication of Rebecca performing the unitizing operation is her response to the teacher's reference to checking. She independently dragged the unit stick to the end of the two-unit stick to see if Tania's new stick was three units long (see Figure 3). Rebecca's action of checking is a confirmation of her understanding that the two copies were the same length and that each of them was the same length as the original unit stick. Acting in this way indicates that she was at a level above the perceptual situation, productively operating with the unit sticks.

Rebecca performed unitizing and uniting operations (Steffe and Cobb, 1988) in the context of joining two and three sticks end-to-end. Her teacher's request to make a stick as long as two unit sticks apparently activated her concept of two as well as the unitizing and uniting operations the concept contained. Later in the session, Rebecca made a five-unit stick, again copying the unit stick and joining the copies together. Apparently, her initial activity had been sufficient to activate her number sequence in the sense that any number word of the sequence symbolized the copying and joining operations involved in making a stick composed of a sequence of unit sticks. But at that moment in the teaching experiment, we were not certain her independently executed actions were a sufficient indication of this inference. The question remained whether the operations symbolized by her number words were available to her in the context of Sticks. We formed the hypothesis that Rebecca, but not Tania, had established a connected number sequence.

In a test of this hypothesis in the next teaching session, the teacher posed a missing-stick situation. With a unit-stick and a six-unit stick already on the screen, he turned to Tania:

Protocol 4 (12 May 1992)

Teacher:	Keep on drawing that (six-unit) stick, Tania, so that the whole stick will be 11 units long. Eleven of these down here (points to the unit stick).
Tania:	(Draws three sticks approximately as long as the unit stick.)
Teacher:	Are you trying to make these (points to the unit stick)?
Tania:	Oh, yeah! (Drags the three drawn sticks into a trash box, draws a stick directly below the six-unit stick of the same length.)
Teacher:	How many do you need, Tania?
Rebecca:	(Spontaneously) draw five of those (the unit stick) and you will have eleven.

Rebecca's comment, "Draw five of those and you will have eleven" illustrates what we mean by symbolized copying and joining actions. It was not necessary for Rebecca to actually copy five more unit sticks and join them to the six-unit stick. Rather, taking the six-unit stick as given, she re-presented the copying and joining actions she would have to carry out to make an 11-unit stick. These re-presented copying and joining actions in turn symbolized the action of continuously drawing a stick beyond the six-unit stick.

Solving the missing-stick situation went beyond the numerical operations Rebecca had performed in the previous teaching session (Protocols 2 & 3). Her solution confirmed that her previous operating constituted a *generalizing assimilation* of her number sequence. To be generalizing, there must be a modification of the assimilating operations. Rebecca's unitizing operation was modified to include the motion of moving the cursor across the screen and the result of this movement, the drawn stick. This modification introduced the property of the length of a stick, the abstracted motion. Moreover, the uniting operation was modified to include the records of the operation of joining sticks end-to-end.

Rebecca's modifications of the assimilating operations of her number sequence constituted an accommodation because they were permanent. The activation of the unitizing and uniting operations constituted the assimilating part of the generalizing assimilation, and their modification constituted the generalizing part. But we would not call an assimilation generalizing unless the operations of her number sequence could be now applied to connected numbers in the same way they could be applied to discrete numbers. That is, operating in the specific situation of making a two-unit stick and a three-unit stick would have to be sufficient to operate in any other situation involving connected numbers that from our point of view was similar to a discrete situation she could solve using her number sequence. This is the observer's meaning of the generalizing part of a generalizing assimilation. It can be accounted for by the identical operations symbolized by any number word of Rebecca's verbal number sequence – the unitizing and progressive uniting operations. Other contextual factors, however, could influence a child's use of his or her numerical schemes. The child's belief, for example, that counting was prohibited, could block a generalizing assimilation like Rebecca's.

A Retrospective Accommodation

We cannot attribute Tania's lack of activation of counting (Protocol 4) to the absence of an interiorized counting scheme, because Tania, like Melissa, could solve missing-item tasks in discrete situations by counting from the numerosity of the part up to the numerosity of the whole. But Tania had yet to establish a scheme for making multiple-unit sticks. Following the situation described in Protocol 4, it was the teacher's goal that she would use the COPY and JOIN commands to make such a multiple-unit stick. However, having no confidence that she could independently operate in that way, the teacher addressed Rebecca first. The hope was that Tania – using her number concepts – would assimilate Rebecca's actions rather than simply imitate them.

   Rebecca made a stick about 1 cm long ("A little eensy weensy one") and then, upon the teacher's request to make the stick twice that long, copied the stick, joined the two pieces, and erased the mark in the middle. The teacher turned to Tania:

Protocol 5 (12 May 1992)

Teacher:	Make a stick that is twice that long, Tania (points to Rebecca's stick).
Tania:	(Draws a stick about 1 cm long, makes two additional copies end-to-end, attempts to erase the marks.)
Rebecca:	You need to join it first.
Tania:	(Joins the three pieces, does not erase the marks.)
Teacher:	(Places the cursor on Rebecca's stick) make a stick that is three of these (changing the language in an attempt to communicate).

Tania:	(Makes four copies of Rebecca's stick.)
Teacher:	Three, not four.
Tania:	(Puts one stick into the trash and joins the three others; does not erase the internal marks).

The teacher then asked Tania to make sticks 6 and 11 times as long as Rebecca's stick. In both tasks, Tania independently made copies of Rebecca's stick and then joined the pieces; again, she did not erase the internal marks. From our perspective, Tania had established an action scheme for making continuous but visually segmented numbers.

Tania's initial imitation of Rebecca's actions is indicated by her choice of a stick as small as Rebecca's original stick and by her omission of the joining operation. The teacher, rather than asking her to analyze what she had just done, took advantage of her actions and of what he believed was obvious to her – that she had made a three-unit stick – and asked her to make a stick as long as three of Rebecca's sticks. Tania's solution to this task can be seen as transitory between her imitation of Rebecca's actions and her independent completion of the 6-unit and 11-unit sticks in that the teacher's intervention ("Three, not four") served to activate her numerical concept of three.

We see the function of imitation in Tania's activation of her concept of three in providing the necessary experience of making a three-unit stick by copying and joining. With her number concepts activated, Tania then constituted the task of making sticks 6 and 11 times as long as the unit stick as one of making sticks composed of a definite numerosity of unit sticks. From the point of view of the actions she had established by imitation, this action scheme arose as result of a retrospective accommodation (Steffe, 1988, 1991c). A retrospective accommodation involves selecting and using conceptual elements already constructed (Steffe, 1991c, p. 42). From the child's side, a retrospective accommodation is self-initiated in that it is the child who must select and use the concept. From an observer's perspective, the conceptual elements may be selected as result of interactive communication, as in Tania's case.

A Functional Metamorphic Accommodation

Tania made further progress during the next task when the teacher asked the students to make a 17-unit stick without using the unit stick. Tania exclaimed, "Oh, I know!" and smiled for the first time in the teaching session. While Rebecca copied the 11-unit and the 6-unit sticks, placing them end-to-end, Tania spontaneously counted-on, 6 beyond 11. She was now very excited, so her teacher asked her to make a stick 23 units long. She copied the 11-unit stick twice and then copied the unit stick onto the end of those two sticks, an indication her number sequence and her adding schemes had been activated.

Tania's insight ("Oh, I know!") indicates her action scheme for making continuous but visually segmented numbers was reconstituted at the level of interiorization. The constraint of not being able to make the 17-unit stick by repeatedly copying and joining the unit stick led to a reorganization of her perceptual field when she regarded the two visually segmented numbers as input for further operating. The two sticks were now agents of activation as well as results of acting. As activating agents, they were reprocessed, and this reprocessing reconstituted them as a situation of Tania's adding scheme, which itself was an interiorized scheme.

The activation of Tania's adding scheme is indicated by her spontaneous action of counting-on. The two sticks, conceptually united, formed a possible stick 17 units long. This uniting action was the result of using the operations of the adding scheme prior to executing its activity – counting on. The restructuring was immediate and it did not involve her subsequent counting activity, which was carried out to confirm the insight.

It might seem unusual for a metamorphic accommodation (Steffe, 1991c) to follow so closely after a retrospective (imitative) accommodation. But there were two elements making the interiorization of her scheme possible. First, the numerical concepts Tania selected and used in making visually segmented numbers were interiorized concepts. Second, her adding scheme operating on her numerical concepts was an interiorized scheme. Thus, it was possible for Tania to raise her action scheme for making continuous but visually segmented numbers to the interiorized level. At this level, the action scheme was reconstituted as a connected number sequence.

A functional accommodation of a scheme occurs in the context of using the scheme. To be metamorphic, a functional accommodation must occur independently, reconstitute the scheme on a new level, and reorganize the scheme on that level. A functional metamorphic accommodation is a reformulation of a strong form of Piaget's reflective abstraction (Piaget, 1980):

Logical-mathematical abstraction ... will be called "reflective" because it proceeds from the subject's actions and operations ... we have two interdependent but distinct processes: that of projection onto a higher plane of what is taken from the lower level, hence a "reflecting", and that of a "reflection" as a reorganization on the new plane. (p. 27)

Reprocessing visually segmented numbers can be viewed as projecting ("lifting") the copying and joining actions to an interiorized level. The reorganization followed from the activation of the adding scheme. By conceptually uniting the two visually segmented numbers into one possible continuous number, Tania could now form a possible stick of 17 units in anticipation.

Functional metamorphic accommodations compliment what has been referred to as developmental metamorphic accommodations (Steffe, 1988, 1991b,c). The basic difference is that, in the functional case, the projecting and reorganizing operations are already available at the level at which the scheme is reorganized, whereas in the developmental case, the projecting operations must be assembled in the experiential situation, and the projection from one level to the next is a protracted process.

We have explained Tania's insight as a functional metamorphic accommodation of her scheme for making continuous but visually segmented numbers. From the perspective of her adding scheme, it would have been possible to account for her modification as a generalizing assimilation. However, this account would ignore the immediate experience of the child. From Tania's perspective, she was making copies of a unit stick and joining them together, so we explained her accommodation in terms of this scheme rather than in terms of her adding scheme. Nevertheless, any accommodation is relative to the operations constituting the accommodation. In the case in which these operations are part of the assimilating operations of another scheme, that other scheme is subject to being generalized as a result of its use in the new situation.

SOCIAL INTERACTION AND SELECTION PRESSURE

The children's production of pluralities in Toys and of continuous units in Sticks provided us with a starting point to transform their initial cognitive play activities into mathematical activities with a playful orientation. Melissa's use of her number sequence, for example, had indeed a playful orientation, carried over from her cognitive play activity.

Although it was our goal to engage the children in mathematical activity, we did not intend to coerce them into accepting our numerical goals. Rather, through our interaction with the students, we wanted them to select numerical goals as if this selection was their idea. We knew they had constructed the numerical schemes to be selected – their number sequences; the relevant conditions for selection were present. Our interactions with them included applying *selection pressure* (Bickhard, 1992). In the case of Melissa and Courtney, the selection pressure took the form of a constraint ("Don't use the mouse"). This constraint minimized the usefulness of producing pluralities and opened the possibility for the children to select counting, that is, their number sequences.

In the microworld Sticks, producing a continuous unit (a stick) did not immediately generate a situation of a child's number sequence. Consequently, a child could not select his or her number sequence as being relevant without modifying the continuous act of drawing a stick into a continuous but segmented action. The necessity of this modification was an opportunity to observe the children's accommodations in their number sequences sufficient for those number sequences to be used in Sticks.

The teacher's initial request to draw a stick as long as two unit sticks (Protocol 2) represents the type of question we believe encouraged the children to select away from cognitive play toward the use of their number sequence without losing the playful orientation. For Rebecca, the almost immediate selection of her number sequence produced an acute sense of power and control not evident in her previous actions in Toys. She was more assertive in her interactions with the teacher and with Tania. For example, she controlled the social interactions and spontaneously verbalized her insights ("Draw five of those (the unit stick) and you will have eleven", Protocol 4). The power of Rebecca's mathematical operations modified her social interactions.

Tania's re-presentation of Rebecca's actions (Protocol 2) was not sufficient to activate her unitizing and uniting operations symbolized by her number sequence. Lacking the independence produced by these operations, she had no choice but to imitate whatever of Rebecca's actions she perceived as successful. We see Rebecca's initial control of the mouse as a possible reason why Tania did not select her numerical operations. In the role of an observer of her partner's actions, Tania was not active enough to independently form and pursue her own goal.

During the teaching session on 12 May 1992 (Protocol 4), Tania seemed to lose confidence. It therefore became critical for the teacher to induce the appropriate actions that would enable Tania to select her number sequence. Returning to the situation in which Rebecca had made her generalizing assimilation and Tania had established her imitative scheme of action, the teacher tailored his interactions with Tania as closely as he could to her language and actions. Nevertheless, the teacher was working blindly in the sense that he could not predict which element of his interaction with Tania might be sufficient to activate her numerical concepts. The key seemed to be the request to make a three-unit stick immediately after Tania had joined three sticks together. The teacher's comment, "Three, not four", served in activating Tania's numerical concept of three, which in turn served to guide the subsequent copying and joining actions. This function of her concept of three for making the three-unit stick was apparent only after she successfully made the 6- and 11-unit sticks.

With Tania's numerical concepts active, the teacher decided to induce further selection pressure by introducing a new constraint: to make a 17-unit stick without copying the unit stick. The goal was for Tania to select away from the copying and joining actions and toward her numerical adding scheme. The teacher achieved his goal because Tania selected her adding scheme. Tania's intense excitement and emotion were an indication of the selection processes at work, processes of which she was totally unaware. In fact, her intense emotion was a side-effect of the selection process, which can be understood when considering that all of her former numerical knowledge became potentially relevant in the current situation and that she had achieved a novel reorganization of her perceptual field.

Phenomenologists like Marton and Neuman (1990) would say that Tania's insight occurred to her in the experienced situation without consciously reflecting on it and that this insight could not be separated from the situation:

Suddenly something stands out as a figure against a background, giving them an intuition of how it might be solved (p. 69).

From Tania's point of view, we completely agree. But this view does not take into account the observer's perspective of the interactive mathematical communication. Without the selection pressures induced through the social interaction, Tania would have made no mathematical progress. Selection pressures might be necessary for children to transform their cognitive play activity into mathematical activity.

INDEPENDENT MATHEMATICAL ACTIVITY

Independent mathematical activity is similar to play in that the teacher's interventions are minimized and the children's spontaneous ways and means of operating are maximized. In contrast to play, which is generally regarded as initiated by children without intervention by an adult, the teacher is involved in the initiation of independent mathematical activity by helping the children establish initial situations and goals. We illustrate this role of the teacher and the emergence of an independent mathematical activity with an example from the teaching session on 15 May 1992 with Tania and Rebecca.

At the beginning of the session, the teacher asked the two children to make a set of sticks ("a staircase"), from one to eight units long. Upon completing this part of the session, the children had a sequence of eight unmarked sticks graduated from one to eight units; a marked eight-unit stick had been placed at the bottom of the screen.

Protocol 6 (15 May 1992)

Teacher: Can you find two sticks that together –

Tania: Equal eight!

Teacher: Good idea!

Tania: (Almost immediately) five and four!

   Teacher: Would that work?

While Tania counted from five to eight to check her initial solution, Rebecca contributed another combination, "One and seven". The children then set out to make the two eight-unit sticks, copying and joining the appropriate unmarked sticks. As the teacher attempted to pose the next problem, Tania interrupted, "There's another way you can make eight, four and four".

We see Tania's completion of the teacher's sentence ("Equal eight") as the initiation of an independent mathematical activity. The situation, the eight unmarked sticks and the marked eight-unit stick together with the teacher's introduction ("two sticks together"), was organized by Tania as an addition situation, as indicated by her anticipation of the teacher's task. She was now in control of the situation and had made the task her own. The teacher's question, "Would that work?" made Tania aware that estimation in this case should be verified. The way was now open for the children to continue on their own, that is, to find the pairs of sticks that would make a stick eight unit sticks long. There was also the possibility for the children to generate a new but similar situation by exchanging the eight-unit stick for a stick of different length and then finding the appropriate pairs that would make the new stick.

The children's repeated use of their adding schemes was accompanied by the internal satisfaction produced by being able to anticipate selecting sticks which, when joined, would make an eight-unit stick. There was also the functional pleasure of using the same scheme repeatedly, of knowing how to operate in a novel situation. The children were engaged in mathematical play, which in this case could be regarded as repeated mathematical experience. Repeated experience, however, is a conservative interpretation of the children's independent mathematical activity. As they made the possible combinations of sticks, they were assembling the material for learning to reason strategically.

In designing the initial situation, the teacher wanted to thematize the children's number sequences and their adding and subtracting schemes across the composite wholes in Toys and the connected numbers in Sticks.

That is, he wanted the children to realize explicitly that they already knew how to operate with connected numbers, a realization that Kieren and Pirie call "folding back" (this volume). Up to this point, their adding and subtracting schemes had been restricted to composite wholes, and the children needed to abstract using these schemes across situations, a process that Piaget called "reflected abstraction" (1980, pp. 27–28). The children were now entering a period in which their independent use of their schemes could reveal uniformity of operating across situations, which is an enlargement of what Cooper has referred to as "repetitive experience" (1991, p. 102), or "practice in mathematical development" (p. 102).

SUMMARY AND FINAL COMMENTS

We have described three types of student activities, as we observed them in the first year of our teaching experiment in the context of the microworlds Toys and Sticks: cognitive and mathematical play, teacher directed mathematical activity, and independent mathematical activity. Figure 4 depicts the relation between the four types of student activities. Initial playful activities such as exploring the possible computer actions and designing elaborate pictures were followed by teacher-directed mathematical activities. The transformation of a situation involving cognitive play into a mathematical situation generally was initiated by a teacher's intervention. On the students' part, this transformation involved assimilations (e.g., Melissa's activation of her number sequence in Toys) and accommodations (e.g., Rebecca's modification of her unitizing and uniting operations in Sticks). During teacher-guided mathematical activities, Tania's path of constructing connected numbers involved imitation and re-presentation of her partner's actions and a retrospective accommodation of her counting scheme. The independent mathematical activities following Rebecca and Tania's construction of connected numbers evolved in a situation posed by the teacher and adopted by the students. This adoption established the students' ownership of the task and was a significant factor in initiating independent mathematical activity. The students' independent mathematical activity can turn into mathematical play, that is, into independent mathematical activity with a playful orientation.

Social interaction among the students of each pair and among teacher and students was a vital component of all activities in the context of the microworlds. The social interaction between teacher and students can be seen as selection pressure; the teacher introduced constraints that encouraged the students to select their mathematical schemes in novel situations.

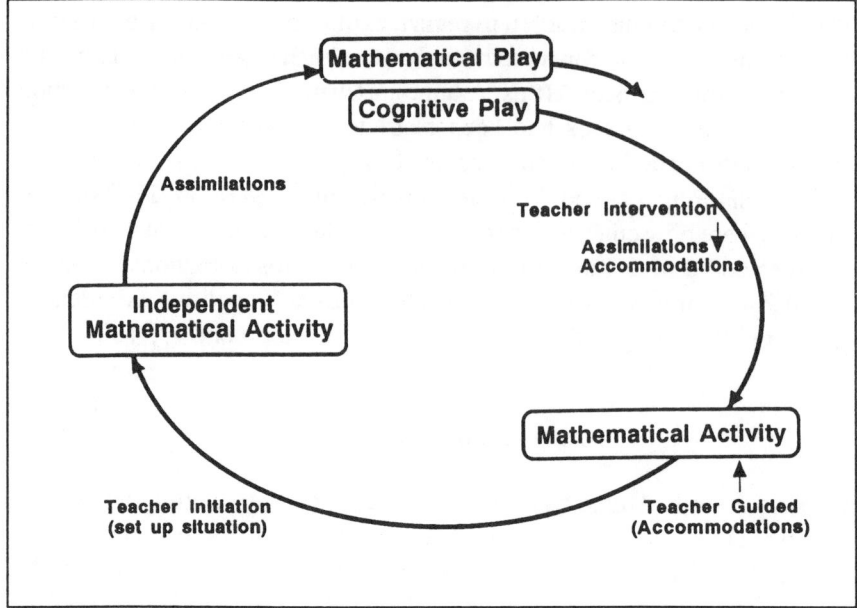

Fig. 4.   Types of student activities.

We wish to emphasize that, within the microworlds, the computer operations were designed to be compatible with our understanding of the mathematical concepts and operations of children, but they were not identical to those concepts and operations. This relation between computer commands and children's operations echoes Piaget's (1980) hypothesis that "the origin of logico-mathematical structures ... cannot be localized either in objects or in the subject" (p. 26). Rather, mathematics is a product of the functioning of mind, a view not leading to any particular specification of situations of learning for children, because such situations do not a priori embed children's mathematical concepts and operations. Our focus is on situations of learning from the point of view of children. Nevertheless, children will not sustain a mathematical activity unless they experience satisfaction in the process of that activity. This satisfaction can emerge from the use of their already available schemes, from operating in a dynamic learning environment, from a successful organization and reorganization of their experience, and from a playful orientation of the activity. The risk of sounding entrepreneurial not withstanding, we found the dynamic aspect of Toys and Sticks to captivate the children's interest. It allowed them to use their schemes in ways not possible when working in a mathematics workbook or with structured learning material, like Cuisenaire rods. From the children's perspective, acting in the microworlds created experiential situations that retreated with every accommodation. That is, as the children imputed structure into the situations through their acting, the situations correspondingly changed and new possibilities for action emerged. In essence, the microworlds opened pathways for mathematical activity.

ACKNOWLEDGEMENTS

The research reported in this paper was conducted as part of the activities of the NSF Project No. RED–8954678 *Children's Construction of the Rational Numbers of Arithmetic*. All opinions expressed are solely those of the authors.

NOTE

[1] We interpret the translation from French as referring to *schemes* rather than to *schemas*.

REFERENCES

Bickhard, M.: 1992, "How does the environment affect the person?", in L. T. Winegar and J. Valsiner (eds.), *Children's Development Within Social Contexts: Metatheoretical, Theoretical, and Methodological Issues*, Lawrence Erlbaum, Hillsdale, NJ, pp. 63–92.

Cobb, P., Wood, T., and Yackel, E.: 1990, "Classrooms as learning environments for teachers and researchers", in R. B. Davis, C. A. Maher, and N. Noddings (eds.), *Constructivist Views on the Teaching and Learning of Mathematics*, National Council of Teachers of Mathematics, Reston, VA, pp. 125–146.

Cooper, R. G. Jr.: 1991, "The role [of] mathematical transformations and practice in mathematical development", in L. P. Steffe (ed.), *Epistemological Foundations of Mathematical Experience*, Springer, New York, pp. 102–123.

Hilgard, E. R. and Bower, G. H.: 1966, *Theories of Learning*, Appleton-Century-Crofts, New York.

Kieren T. E. and Pirie S.: this volume, "Growth in mathematical understanding: How can we characterize it and how can we represent it?" *Educational Studies in Mathematics*.

Marton, F. and Neuman D.T: 1990, "Constructivism, phenomenology, and the origin of arithmetic skills", in L. P. Steffe and T. Wood (eds.), *Transforming Children's Mathematics Education: International Perspectives*, Lawrence Erlbaum Associates, Hillsdale, NJ, pp. 62–75.

Papert, S: 1980, *Mindstorms: Children, Computers, and Powerful Ideas*, Basic Books, New York.

Piaget, J.: 1962, *Play, Dreams and Imitation in Childhood* (Translated by C. Gattegno and F. M. Hodgson), W. W. Norton & Company, New York.

Piaget, J.: 1980, "The psychogenesis of knowledge and its epistemological significance", in M. Piattelli-Palmarini (ed.), *Language and Learning: The Debate Between Jean Piaget and Noam Chomsky*, Harvard University Press, Cambridge, MA, pp. 23–34.

Sinclair, H.: 1990, "Learning: The interactive recreation of knowledge", in L. P. Steffe and T. Wood (eds.), *Transforming Children's Mathematics Education: International Perspectives*, Lawrence Erlbaum Associates, Hillsdale, NJ, pp. 19–29.

Steffe, L. P.: 1988, "Modifications of the counting scheme", in L. P. Steffe and P. Cobb, *Construction of Arithmetical Meanings and Strategies*, Springer, New York, pp. 284–322.

Steffe, L. P.: 1991a, "Operations that generate quantity", *Learning and Individual Differences* 3(1), 61–82.

Steffe, L. P.: 1991b, "The constructivist teaching experiment: Illustrations and implications", in E. von Glasersfeld (ed.), *Radical Constructivism in Mathematics Education*, Kluwer Academic Publishers, Boston, MA., pp. 177–194.

Steffe, L. P.: 1991c, "The learning paradox: a plausible counterexample", in L. P. Steffe (ed.), *Epistemological Foundations of Mathematical Experience*, Springer, New York, pp. 26–44.

Steffe, L. P.: 1992, "Schemes of action and operation involving composite units", *Learning and Individual Differences* 4(3), 259–309.

Steffe L. P. and Cobb, P.: 1988, *Construction of Arithmetical Meanings and Strategies*, Springer, New York.

Steffe, L. P. and von Glasersfeld, E.: 1985, *Child Generated Multiplying and Dividing Schemes: A Teaching Experiment*. NSF Grant No. MD–8550463.

Steffe, L. P., von Glasersfeld, E., Richards, J., and Cobb, P.: 1983, *Children's Counting Types: Philosophy, Theory, and Application*, Praeger, New York.

Thompson, P.: 1991, "To experience is to conceptualize: A discussion of epistemology and mathematical experience", in L. P. Steffe (ed.), *Epistemological Foundations of Mathematical Experience*, Springer, New York, pp. 260–281.

Von Glasersfeld, E.: 1981, "An attentional model for the conceptual construction of units and number". *Journal for Research in Mathematics Education* 12(2), 83–94.

Von Glasersfeld, E.: 1983, "Learning as constructive activity", in J. C. Bergeron and N. Herscovics (eds.), *Proceedings of the Fifth Annual Meeting of PME-NA*, Université de Montreal, Montreal, Canada, pp. 41–63.

Von Glasersfeld, E.: 1989, "Constructivism in education", in T. Husen and N. Postlethwaite (eds.), *The International Encyclopedia of Education*, Pergamon Press, Oxford, pp. 162–163.

Department of Mathematics Education
University of Georgia
Athens, Georgia 30602
U.S.A.

JERE CONFREY AND ERICK SMITH

EXPONENTIAL FUNCTIONS, RATES OF CHANGE, AND THE MULTIPLICATIVE UNIT

ABSTRACT. Conventional treatments of functions start by building a rule of correspondence between x-values and y-values, typically by creating an equation of the form $y = f(x)$. We call this a *correspondence* approach to functions. However, in our work with students we have found that a *covariational* appraoch is often more powerful, where students working in a problem situation first fill down a table column with x-values, typically by adding 1, then fill down a y-column through an operation they construct within the problem context. Such an approach has the benefit of emphasizing rate-of-change. It also raises the question of what it is that we want to cal 'rate' across different functional situations. We make two initial conjectures, first that a rate can be initially understood as a *unit per unit* comparison and second that a unit is the *invariant relationship between a successor and its predecessor*. Based on these conjectures we describe a variety of multiplicative units, then propose three ways of understanding rate of change in relation to exponential functions. Finally we argue that rate is different than ratio and that an integrated understanding of rate is built from multiple concepts.

INTRODUCTION

Constructivism is widely used to support reform efforts in mathematics education. In the community of mathematics educators, significant attention has been paid to its radical implications for student learning. Children, and for that matter, all students, are recognized not simply as inadequate or incomplete adults, but as thinkers and sense-makers in their own frames of reference. Constructivists have effectively documented that student errors are seldom random or capricious – they have a rationality and functionality of their own. In this regard, constructivists have documented that teachers and researchers must pay close attention to how a mathematics problem is conceptualized, worked on and evaluated by students.

What has been given less attention is that the constructivist research program implies much more than revising our views of students and their learning. Our own understanding of the mathematics needs to be challenged and reformulated in light of students' approaches and methods. All too often, the constructivist agenda is reduced to the question of how to get the rich and diverse concepts and strategies of students to be aligned with our own more expert views, and those expert views are assumed as complete, well-structured and non-problematic. One result of such an interpretation of a constructivist program is that reform efforts which attempt to open up and rethink the mathematical content are targeted mostly at the elementary grade levels, while secondary educational reform is more typically limited to pedagogical approaches as the content is assumed to be well-secured in its expert structure. We would claim that we witness too little attention to how

Educational Studies in Mathematics **26**: 135–164, 1994.
© 1994 *Kluwer Academic Publishers. Printed in the Netherlands.*

the research on student conceptions demands that we rethink what we claim as mathematical knowledge across *all* grade levels.

In addition, the increasing recognition that a teaching experiment, in which a researcher and a student interact around a set of tasks, is itself a social interaction implies that results need to be interpreted as a particular form of discourse. The interviewer is trying to understand both the student's problematic and what s/he is doing and attempting to do to resolve that problematic. Simultaneously, the student is constructing a problematic in conjunction with what s/he views as the interviewer's expectations and expects that the interviewer will be more expert in the tasks in which s/he then engages. The result is that when the interviewer analyzes the data to build a model of what the student's actions, words, schemes and concepts are like, the interviewer's own understandig of the task influences that model. What has often been neglected in the past is that what the interviewer learns about the student (as s/he constructs a model of the student) evolves together with her/his own changing understanding of the subject matter. Because of this inseparable and simultaneous development, researchers have tended to describe these changes only in terms of the models of the student and in doing so have neglected their own learning.

In previous work, Confrey (in press a) has labeled this as a distinction between voice and perspective. Voice is used to describe the model of the child – and interviewing is an attempt to give voice to a child's ways of approaching a problem. Perspective surrounds voice, as the context from which the interviewer hears, probes and develops and sets new tasks. In analyzing the data, Confrey suggested that researchers need to try to separate out these two different strands and to acknowledge the extent to which the researcher's own perspective changes – one's own mathematical knowledge evolves through interactions with the student and through the analysis of student voice.

This paper provides an example of how researchers' mathematical knowledge needs to be examined in light of research on students' conceptions. We extend the analysis of perspective even further than in Confrey's previous work to argue that analyzing student data can lead to a need to reexamine and extend our mathematical understanding *beyond the immediate reaches of student voice* – into new territory. This has led us to a whole new territory of mathematical meanings for functions, units and rates.

Although the *National Standards for Curriculum and Evaluation* (NCTM, 1989) advocates placing the topic of functions in a central position in the secondary curriculum, it does not explicitly address the epistemological issues raised as the treatment of functions moves away from an emphasis on formal definitions. In our own work, we have found that as students generate functional relationships by acting within contextual situations and by using multiple representations in both creating and representing their solution processes, legitimate and diverse ways of thinking about functions are created. In this process the meaning associated with the function concept is created through the interaction of context, multiple representational forms, and technological tools (Confrey, 1991a, 1991c, 1992; Rizzuti, 1991; Rizzuti and Confrey, 1988; Smith and Confrey, 1992; Confrey *et al.*, 1991; Afamasaga-Fuata'i, 1991; Borba, 1993; Borba and Confrey, 1992). This research has led us to challenge many attempts to label alternative students' conceptions as either misconceptions or as "prefunctional" (Dubinsky and Harel, 1992). We see this as too often indicative of a narrow view of the function concept, bounded by its placement in a decontextualized and heavily abstract view of mathematical systems. Such a view ignores the richness and viability in the student conceptions and thus lacks serious attention to the epistemic content of the student *voice* and the resulting implications for the researchers' own *perspective*.

In this paper, we have chosen the issue of rate of change as an entry to thinking about functions. Recent research has shown that even young children can use rate of change as a way to explore functional understanding (Nemirovsky and Rubin, 1991; Confrey, 1993; Kaput and West, in press). Arguing against the delaying of the introduction of rate of change issues until calculus, where it becomes studied as a 'property' of functions, we have shown that students exhibit strong intuitive understandings of change and can use this understanding to generate functional relationships. This rate-of-change approach is also closely related to a "covariational" understanding of functions which has both historical roots (Boyer, 1968; Smith and Confrey, in press) and is closely related to student problem-solving actions (Confrey *et al.*, 1991; Rizzuti, 1991). Thus, the balance of this paper will begin with a discussion of covariation then move to examples from linear and exponential functions, discussing our initial discoveries about the exponential function in terms of its underlying operational character and its multiple meanings for rates of change. By analyzing these different meanings, we offer a new conceptualization of the unit and rate concept and articulate the role of the 'split' in rate of change.

Covariation as an Alternative Function Concept

We have described two general approaches to creating and conceptualizing functional relationships, a correspondence approach and a covariation approach (Confrey *et al.*, 1991). Most common in the curriculum is the correspondence approach in which one initially builds a rule that allows one to determine a unique y-value from any given x-value. Thus one builds a correspondence between x and y. This approach is emphasized in conventional algebraic notation in which we write:

$$y = f(x).$$

A covariation approach, on the other hand, entails being able to move operationally from y_m to y_{m+1} coordinating with movement from x_m to x_{m+1}. For tables, it involves the coordination of the variation in two or more columns as one moves down (or up) the table.[1] In working with many applications, students find a covariation approach easier and more intuitive. It can also lead nicely to an algebraic coding of the correspondence rule for a function; however, this skill is learned through a variety of experiences. Supporting the distinctiveness of a covariation vs. correspondence approach, we find that even those well-schooled in building correspondence relations can face a challenge in moving from covariation to correspondence approaches. For example, in an informal meeting, a group of secondary mathematics teachers were quick to approach to a tuition problem (tuition increases at a rate of 11% per year) as covariation, but hypothesized that the correspondence equation would be a polynomial rather than an exponential.

RATE OF CHANGE

A covariation approach to functions makes the rate of change concept more visible and at the same time, more critical. Rizzuti (1991) documents that students find it useful to describe a function in terms of its rate of change, both graphically and numerically. She further illustrates that when students were given arithmetic sequences and geometric sequences, each mapped onto the natural numbers, and asked to describe their graphs, they discussed the differences in terms of rates of change even though they had not been introduced to this language. One student explains the difference in relation to the numeric values:

> Umm..... maybe its because with geometric you multiply, so your numbers would – like you would go from, for example, -2 to -8 to -32. There's so much of a gap between the numbers. Whereas in an arithmetic [sequences] there's gonna be the same amount of gap between the numbers. [p. 126]

Not all students argue as this one did that a geometric sequence has a changing rate. Some students describe the rate of change of the exponential function as constant, consistent with the argument by Confrey (in press b) that a multiplicative rate of change could be regarded as constant. Other students show evidence of multiple views of the rate of change construct as illustrated in the following example. Given tabular data on the growth of bacteria over time, students were asked why the numbers were so large. One student replied:

> Because it's just increasing so rapidly. It's increasing by 16 each – it starts out with three. But just every minute, each of them increasing by 16 – it divides itself up. It makes itself into 16 more cells. [p. 134]

This data begins to suggest the multiple meanings of the term 'rate'. We discuss this further in the following section.

Rate of Change in the Exponential Function

Our early work on exponential functions has led us to consider rate of change as a primary point of entry when working with students. This analysis was spurred by our research on students' development of exponential functions wherein we witnessed:

1.   A student building a number line which combined primary units constructed multiplicatively as powers of ten with subunits increasing additively between each pair of primary units (Confrey, 1991c). This helped us begin to recognize the viability of a multiplicative unit.

t	c
time (hrs)	number of cells
0.00	9.00
1.00	81.00
2.00	729.00
3.00	6561.00
4.00	59049.00
5.00	531441.00
6.00	4782969.00
7.00	43046721.00
8.00	387420489.00
9.00	3486784401.00

Fig. 1. Growing cells.

2.   A student working on a problem where the number of robots increases exponentially. He described the pattern in a table of time vs. the number of robots in the following manner. "If the time interval is multiplied by six then the number of robots produced would be raised to the sixth power" (Confrey, in press a). This allowed us to recognize his ability to coordinate two 'covarying' worlds, one built additively (where the unit is repeated addition) and one built multiplicatively (where the unit is repeated multiplication).

We point out that our discussion of rate will rely on its use in a variety of contexts. Thus, whereas we do not seek a context-independent interpretation of rate, we do seek a description of rate which can describe its use across various contexts.

We often begin teaching exponential functions with the following situation: The number of cells growing in a lab experiment has been recorded every hour. We show three entries of the table presented in Fig. 1. We ask: 'What would you predict for the number of cells present at the next reading?' We use this example to anticipate the question of what might happen at a half-hour, an issue of interpolation. This problem was designed to be discrete, while being amiable to determining a value for the number of cells at the half-interval. Splitting in biology happens most often in two's but does occur in three's and, hence, if sampled every other split, it happens in nine's.

A conventional analysis of rate of change in this problem would involve a straightforward calculation, taking differences in the 'c' column for each change of one in the 't' column. However, in the context of cell splitting, many students chose different ways to describe rates of change. Three such approaches include:

t	c	Δc	Δ(Δc)	Δ(Δ(Δc))
time(hrs)	number of cells			
0.0	9.0			
1.0	81.0	> 72.0		
2.0	729.0	> 648.0	> 576.0	
3.0	6561.0	> 5832.0	> 5184.0	> 4608.0
4.0	59049.0	> 52488.0	> 46656.0	> 41472.0
5.0	531441.0	> 472392.0	> 419904.0	> 373248.0
6.0	4782969.0	> 4251528.0	> 3779136.0	> 3359232.0
7.0	43046721.0	> 38263752.0	> 34012224.0	> 30233088.0
8.0	387420489.0	> 344373768.0	> 306110016.0	> 272097792.0
9.0	3486784401.0	> 3099363912.0	> 2754990144.0	> 2448880128.0

Fig. 2. Differences for an exponential function.

1. *An additive rate of change.* Students will often start by calculating the differences between succeeding values of c. An additive rate of change is defined by finding the differences in succeeding c-values for constant unit-changes in the t-values. Typically they arrange their first column in incremental units of one, avoiding (or possibly neglecting) the unit change issue. By choosing the difference command, students can have Function Probe calculate differences for any column of values. Note that in this problem (Fig. 2), differences increase rapidly as one moves down the table. Also note that taking differences of differences, and differences of differences of differences continues to produce vertically increasing gaps between consecutive values. If, however, one looked at the ratio of these differences (see Fig. 3b), one would find the sequence of the differences to be exponential, a suggestion of what comes when one examines the derivative of an exponential. This pattern in the differences is unlike the situation students have experienced with polynomial functions where for a polynomial of degree n, the nth difference column shows constant values. (See Confrey, 1992, and Afamasaga-Fuata'i, 1991, for a discussion of a rate of change approach to quadratic functions.)

2. *A multiplicative rate of change.* Looking at the first two or three values, students will often suggest that the table is increasing by multiplying by nine, by 'a constant rate of nine'. This suggests a second concept of rate of change which is found by calculating the ratios between succeeding y-values for constant unit-change in the t-values. Students can check such a hypothesis by using the ratio command in Function Probe (see Fig. 3a). In this case, the rate of change is a constant ratio of nine, i.e., nine times as many cells per hour. Because we want to encourage students to develop a strong intuitive sense of ratio change as well as difference change, we have developed a symbol for multiplicative rate of change, ℝ and encourage students to use both ℝx and Δx in their mathematical explorations.

t	c	⊗c
time(hrs)	number of cells	
0.0	9.0	› 9.0
1.0	81.0	› 9.0
2.0	729.0	› 9.0
3.0	6561.0	› 9.0
4.0	59049.0	› 9.0
5.0	531441.0	› 9.0
6.0	4782969.0	› 9.0
7.0	43046721.0	› 9.0
8.0	387420489.0	› 9.0
9.0	3486784401.0	

Δ(Δ(Δc))	⊗(Δ(Δ(Δc)))
› 4608.0	› 9.0
› 41472.0	› 9.0
› 373248.0	› 9.0
› 3359232.0	› 9.0
› 30233088.0	› 9.0
› 272097792.0	› 9.0
›2448880128.0	

Fig. 3.   Ratios for an exponential (a) and for differences (b).

3. A 'proportional new to old' rate of change. We have also had students describe the pattern in this table as an 'increase by eight times' for each change in time by one hour. They describe the amount being added on each time as being eight times the last value. For example in the table of first additive differences (Fig. 2), the amount being added to 9 is 72, which is eight times nine; the amount being added to 81 is 648, eight times 81, etc. This is a way of describing what is new, what is being added, as a proportion of what is old, what was there before.

Many of us might think of this as an awkward way to describe the pattern in this context where the multiplicative rate is an integer. However, its usefulness becomes more apparent in the context of interest rates and compound interest. An interest rate of 5% per year means that the new money will be 5% of the old money and then must be added on to get the total amount in the account. This use of rate sits squarely between an additive and multiplicative treatment of rate, and can cause ambiguities in the use of language. For example, if one says that one's principle has increased 4% per year, one would assume that $100 would become $104 after one year. If one's principle has increased 200% per year, then one could mean that $100 becomes $200 (multiplicative rate) or $100 becomes $300 (proportional new to old) after one year. In compound interest, the formula for the final principal is $P_f = P_i(1 + r)^t$. The rate here is the r in the formula or 4%/year in the first example.[2]

What is a Rate?

These three alternative approaches to change in the exponential led us to claim that the treatment of rate in many mathematics textbooks and classrooms is ambiguous and overly narrow. Our epistemological investigation led us to argue that the dominance of calculus in the mathematics curriculum encourages instructors to emphasize an additive view of rate while neglecting the legitimacy of the multiplicative approach. Our intention is to remedy this situation by building a broader analytic approach to rate of change as follows: rate is a unit per unit comparison. This analytic approach will be embedded in a qualitative approach to rate and in an epistemology of multiple representations. Before doing so, we need to discuss our approach to the concept of a unit.

A unit is "the invariant relationship between a successor and its predecessor; the unit is [created as the result of] the repeated action" involved in numeration (Confrey, in press b). For example, in order to create a *counting* unit of one, a child must first recognize a multitude composed of objects sharing a particular quality. For young children the objects are typically identical, varying only in location or color, like three fishes. The child then points a finger at each object in turn and utters a sequence of number words. The unit which s/he creates is the result of the operation of carrying out mentally the repeated action (which may or may not be done physically by gestures) and forming it into a unit. Thus, adding one, the primitive that later is modified to become addition of $n > 1$, emerges concurrently with the unitizing activity, the count. This is a claim that the genesis of numbers and basic unitizing operations are inextricably intertwined.

As Steffe has demonstrated, creating units out of non-one groups involves a complex developmental sequence of making the units, treating the composite as a unit and iterating it, and finally creating an anticipatory scheme that allows one to imagine mental actions of carrying out the unitizing operations (Steffe, 1990, in press). In our work, we seek to broaden the definition of unit to include other underlying repeated actions, particularly the action of splitting (Confrey, in press b). However, before we examine the splitting construct, we wish to discuss the implications of defining a unit in terms of the mental operation of internalizing a repeatable action, rather than in terms of standardized units such as inches, pounds, etc. This approach to *unit* has a number of radical implications:

1.   It stresses that units are not 'things' that can be disassociated from the mental operations involved in their construction. That which one identifies as the *same* (*unit*-izing) in these repeated actions evolves into the unit. As a mental operation, unitizing frequently involves situations into which a person enters with an intention or goal. The goal might be to answer the question 'How many?' or 'How much is there?', in which case the answer is intended to be a number. However, units can also relate to non-numeric goals as designing and cutting out swatches to make a quilt (see Fig. 4). The unit encapsulates what the actor perceives as repeated action. This involves a mental segmentation of a potential or real physical action. Our approach does not ignore standard unit-names (inches, lbs, etc.), but recognizes that these are social conventions. As these are standardized onto measurement tools, rulers, clocks, calendars, we tend to separate the units from their construction; however, watching children learn to use a ruler will quickly convince one of the process of the development of mental operations involved. Recall that feet were originally connected to the footstep and meters to the distance between the nose and the hand. As these standard unit-names become integrated into our construction

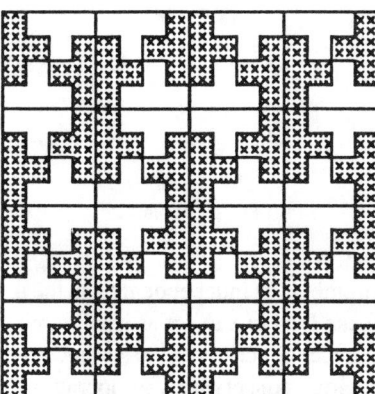

Fig. 4.   A quilting pattern.

Fig. 5.   Possible quilting units.

of units, they come to play a role in the units we construct.

2.   It means that units involve mental operations for *comprehending change or pattern*. That is, we come to perceive pattern and/or change by answering two questions simultaneously: 'What is different?' and 'What is the same?' As we strive to answer these questions, we create the pattern or unit via *segmentation*. Sometimes an obvious choice of unit exists, such as in discrete cases when we count, for example, the children in a classroom. In others, we might override the discrete choice in favor of a unit that is perceived to be more relevant to our goal, such as when we buy potatoes by the pound rather than by the potato (to assure price uniformity over potatoes of varying sizes). Other times, such as in a rainbow, the segmentation is arbitrary and conventional. Does a rainbow always have eight colors? Consider the pattern in Fig. 4 (from Rubin, 1992). Depending on how one views it, it can be described as having multiple candidates for the unit, and each candidate leads towards different actions in creating the quilt. These units could include those presented in Fig. 5.

3.   It implies that initially it is the 'unitizing' operations that create numbers. That is, it is the goal of finding out how many or how much and the operation of creating units that must proceed before numbers are available as an answer. That answer, represented as a number, is *made* from the units. Two errors are frequently made if one does not recognize this claim:

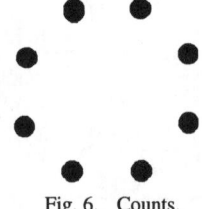

Fig. 6. Counts.

a. First, treating numbers as independent from the unitizing operations used to create them can lead one to act as if all numbers are the same kind of entity. For instance, Greer (1988) has argued that we should encourage students to develop "conservation of operation" so that the operation is seen as independent of the number type. We would claim that numbers have "biographies" and that their character is inherently tied to how they are constructed. Thus we would view a claim for an interaction between a number type (ratio, decimal, fraction, integral) and an operation as a reasonable expectation on the part of students.

b. Second, treating numbers as logically prior to the operations leads us to assume that all numbers are characteristically the same, regardless of how they are made. Thus the number distinctions we do make (counting, integer, rational, real) are only loosely related to the operational bases for the numbers and are, to a large extent, a posteriori distinctions. For example, we neglect the geometric origins of square and cube roots treating them simply as unending decimal strings like all other irrationals. Likewise, we call 0.2, 1/5, and $\sqrt{1/25}$ the *same*, despite the differences one might expect in their constructions implied by the notations.

4. It can lead one to make certain assumptions about numbers while failing to recognize that assumptions cannot be separated from the operations used to make those numbers. For example, one counts to create the natural numbers and then the natural numbers have a unit of one, symbolizing the underlying repeated one-count. When we see a display such as the one shown in Fig. 6, we are likely to say immediately that there are eight objects. However, if someone then tells us that the figure is a schematic for planning the steps in a square dance, we look again and suddenly see four objects (couples). It is that which we make the same as we count around the circle (or square) that makes the unit and simultaneously creates that number.

Additive Units

To illustrate our treatment of units, we will work through an example. Suppose a house is located at a certain distance away from a cliff which is eroding towards the house. For five years, the person measures the distance to the cliff and the table is shown in Fig. 7. Since she wants to predict when the house will fall over the cliff, she analyzes her table for the amount of change in the distance each year.

t	d	Δd
time (yrs)	distance (inches)	units
0.00	100.00	
1.00	96.50	> −3.50
2.00	93.00	> −3.50
3.00	89.50	> −3.50
4.00	86.00	> −3.50
5.00	82.50	> −3.50

Fig. 7. An additive unit.

The unit as rope length

t	d	Δd
time (yrs)	distance (rope lengths)	units
0.00	28.50	
1.00	27.50	> −1.00
2.00	26.50	> −1.00
3.00	25.50	> −1.00
4.00	24.50	> −1.00
5.00	23.50	> −1.00

Fig. 8. An additive unit.

According to our units analysis, the repeated action in the example is the annual loss of 3.5 inches. Thus, we claim, in this example, the unit is −3.5 inches. Many would be inclined to identify inches as the unit. Inches do describe the repeated unit used in measuring a single 3.5, as the ruler is slid along to mark each inch, and as such they are units, but they are not units to be used in describing the experiential invariance in the erosion of the cliff. The decision to segment one's experience into annual intervals sets the stage for a unitizing operation to create a unit of −3.5 inches from the initial one inch measure. Thus, a unit is constructed in relation to an experiential situation which involves both segmentation and an invariance across segments. In this case, this invariance could be described as -0.29 feet or as −8.9 cm. Likewise, the owner could have constructed an equally viable unit by stretching a rope from the house to the cliff edge and marking off the annual changes in the rope length. This action would create the unit shown in Fig. 8. If she used that unit to measure the original distance and made a table equivalent to the one above, she could have created the table shown in Fig. 8.

Our selection of a standard unit name (inches) to describe the unit of change in a situation is one of convenience, efficiency and the form of measurement tool. The lesson here is that a unit is defined in relation to the situation and to the recognition of a repetition. In relation to these situations, we use unitizing operations similar to those described by Steffe to create a unit from standard unit names. However, the sense of invariance across units is not necessarily tied to the selection of the unit name. Communicating to others is simply made easier by the use of standard units.

What is particularly pleasing about such an approach is that if a student looks at this example in terms of its unit (-3.5 inches), is aware that this unit is created by segmenting the change over a year, and wants to create a correspondence relationship between the number of years and the the the distance to the cliff, s/he may recognize that it will be:

{the initial distance to the cliff} plus {the number of years} $*$ {the number of units}

and write:

$$d = 100 + t * (-3.5).$$

That is, we suggest that we might revise the traditional form of $y = mx + b$ to read $y = b + x(m)$ where m is the unit of change. That is, take an initial value and increment (or decrement) it by x units of change. We have found evidence that students often do, in fact, approach linear problems in this way, particularly when using the calculator on Function Probe (Confrey *et al.*, 1991b).

Multiplicative Units

This treatment of the unit creates an interesting possibility. Suppose one begins the process of quantification with a different action than counting. Could another equally legitimate unit emerge that would open up a world of quantitative possibility that remains hidden without it? This idea is the basis of the 'splitting conjecture':

> Among the actions we observe in children and which can be interpreted multiplicatively are some that are related to addition (affixing, joining, annexing, removing, etc.) and others that seem relatively independent of addition (sharing, folding, dividing symmetrically, and magnifying). It is in these latter actions that we find the basis for splitting. In its most primitive form, splitting is defined as an action of creating simultaneously multiple versions of an original, an action which is often represented by a tree diagram. As opposed to additive situations, where the change is determined through identification of a unit then counting consecutively copies of that unit, the focus in splitting is on the one-to-many action.... Closely related to this primitive concept are actions of sharing, where verifying the outcome can be based on looking for congruence of parts, rather than counting... Splitting can also be differentiated from counting (and repeated addition) by its geometric connections to similarity. From this it can be seen that similarity forms the basis for our depth perception as we maintain the identity of objects as they move towards (magnification) or away (shrinkage) from us (Confrey, 1990, pp. 1–2).

Thinking of splitting in terms of the above description of units, a repeated splitting action should lead to the creation of a *multiplicative unit*. Consider, for example, the following situation in which a child shares a pie: a child splits a pie in half, and then splits the halves in half. In carrying out the action again, would she do one additional cut (resulting in two quarters and four eighths) or two additional cuts resulting in eight eights? In our experiments with children, they typically make two additional cuts. Their reasoning is that they want to have equal pieces, and to do so, they must cut each piece in half. The repeated action cannot be simply the physical action of cutting, but must be the action of cutting each piece into two equal parts. We would describe this as a three 2-splits, or three multiplicative units of two.[3]

In the cell example given earlier, there are nine times as many cells every hour. The invariance between predecessor and successor is most immediately described as 'nine times as many'. This is the unit. Notice we did not include the word 'cells' in the description. 'Cell' is not a description of the repeated action which created the unit in this situation. As opposed to the cliff problem, where the additive unit, -3.5 inches, could always be made from iterating 1 inch 3.5 times (viewed as taking away), one cannot create the splitting unit, '9 times as many' from 'cells' for there is no invariant relationship between '9 times as many' and 'cells'.[4] In fact, if the number of cells was changed, for example, to the number of 'pairs' of cells (or number of 'hundreds', or number of 'millions', etc.), the invariance would still be 'nine times as many' each time. That is because the change between number of cells and number of pairs of cells involves counting in an additive world. Thus the standard unit names from the counting world will *not* be relevant to our discussion of units in the splitting world.

There are, however, standard units in the splitting world which will make a difference in how we describe multiplicative units. They do not look like our conventional additive standard units *at first glance*. However, by creating a language to describe them, they create an appropriate measuring device for converting among units in a splitting world. Let's begin by creating a 'state' type name for the unit nine times as many.[5] Call it a nine-split. Now, suppose that in the cells example, the researcher observes the cells through a microscope and sees that every time an individual cell splits, it actually splits into three new cells. This observation may change the perspective she chooses to take. If she becomes interested in the 3-split from a biological perspective, she may choose to identify this as a unit and perhaps to create a table such as the one in Fig. 9. Although she may recognize that this repeated action will occur (on the average) at half hour intervals (and thus that the first column could be labeled 'Number of half hours'), her focus in creating the unit of a 3-split was a change from segmenting a unit in time (every hour) to segmenting a unit in reference to a biological process.

If the researcher maintains her interest in describing the repeated action each hour, the nine-split is still the unit of change. She may choose, however, to describe it as two 3-splits. The unit is created through her intention to seek out and organize her experience in terms of a repeatable action.

Within this multiplicative framework, it would be relatively basic to assert that the amount of time required for a quantity to double would be fixed over the life of the biological process. This is because the action of doubling creates a multiplicative unit which is then paired with an additive unit in time. Doubling (time) and halving (half-life) are standard multiplicative units in their own right

n	c	$u = \circledR c$
number of 3-splits	number of units	unit
0.00	9.00	> 3.00
1.00	27.00	> 3.00
2.00	81.00	> 3.00
3.00	243.00	> 3.00
4.00	729.00	> 3.00
5.00	2187.00	> 3.00
6.00	6561.00	

Fig. 9.　A 3-split unit.

and become part of the fabric of the exponential function within such a perspective.

To convince one that this use of multiplicative units is not as strange as it might seem, consider the following situation. A child is asked how many biological great-great-grandparents she has. She answers '16' and says 'I doubled four times': 2 parents, 4 grandparents, 8 great-grandparents and 16 great-great-grandparents. The number of great-great-grandparents is 4 generations back of *doublings* for the little girl. A 16-split is four 2-splits. We propose that doubling and halving are the most primitive of units in a splitting world. We further propose that there are relatively few familiar primitive units in the sense of doubling and tripling. As argued before, 2-splits, 3-splits, 5-splits, 7-splits, 10-splits (often created as a 5-split followed by a 2-split in harmony with our fingers or toes), cover most splits we use regularly, i.e. most of what we would call standard multiplicative units (Confrey, 1990, in press b).

The different units in the splitting world may appear to be dull in comparison with the counting world with its linear measurements: feet, inches, centimeters, miles, etc.; its volume measures: cups, cubic cm; its area measures: square inches, acres; and its weights: tons, pounds, grams, and so on. Because of the more common usage of additive units, the vocabulary for describing them is more robust and widely recognized; however, we would also suggest that as we learn to be more discriminating in our visualization and articulation of multiplicative units, we will begin to make more distinctions and create more diversity within multiplicative worlds. The standard additive units we use can often be classified in terms of the situational action we take in making them. Likewise, in multiplicative situations, we have situational actions which serve to classify multiplicative units. Parallel to counting, in discrete situations, we have used splitting-units. We also might wish to distinguish sharing-units, similarity-units, percentage-units, magnification-units, order-of-magnitude-units, and probability-units.[6] Some examples are given here.

1. *Splitting.* The split (most easily viewed in discrete cases) is easily represented by the tree diagram and involves replication (see Fig. 10). The repeated action is one of an object being replaced by some fixed number of copies as a repeatable process. Thus we measure it in copies, and answer the question, 'so many times

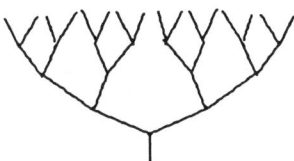

Fig. 10.   Splitting.

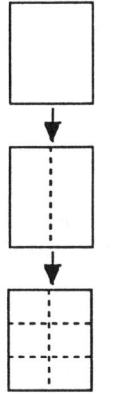

Fig. 11.   Sharing.

as many'. Both cell splitting and the creation of family trees represent the use of splitting units.

2. *Sharing.* A sharing-unit, such as the one illustrated in Fig. 11, is a unit created with the intention of creating a certain number of equal shares, an action common in young children and, for which, they develop a variety of strategies (Pothier and Sawada, 1983; Confrey, 1992). Confrey found that elementary students were able to develop a variety of strategies for sharing cookies with friends which included both basic shares (or splits) into two or three pieces and compound shares such as sharing with 4 by two consecutive 2-shares or with six by a 2-share followed by a 3-share. Since sharing is undertaken with the intention of creating a certain number of shares, we would call a 2-share an intention to share with two such as dividing a cookie in half or creating 2 equal size piles of candy from one original pile.

3. *Similarity.* In situations where we identify growth or change through the identification of similarity in shape, and are further able to segment this change in such a way that a constant proportionality is maintained across segments, we identify a similarity unit. Although many organisms grow in such a way that they maintain a similar shape, we do not necessarily identify a similarity unit because we have

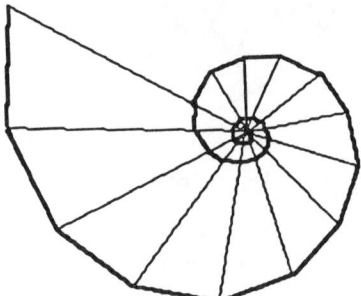

Fig. 12.   Proportional similarity.

no apparent segmentable action that maintains a constant proportion across units. However, in cases like the nautilus shell, which grow by periodically creating new chambers which are similar and maintain a continuing constant proportion, we can identify a similarity unit. The growing triangles in Fig. 12 illustrate this pattern.

4. *Percent.* The multiplicative unit with which we are probably most familiar is percent. In its simplest form, percent is a standardized way of indicating an action based on a proportionality: 'percent is to 100' is the same as 'new is to old'. Perhaps because of its common use, percent has developed several context-specific idiosyncrasies, especially when used as a unit in a rate relationship. Davis (1988) posed the question: 'Is percent a number?' We would answer that percent is a multiplicative unit.

5. *Magnification.* Although magnification as a multiplicative unit is similar to proportion/similarity, magnification is generally used to indicate segmentation relative to different perspectives on the same object whereas proportional similarity is used to indicate a relationship between separate objects, such as the chambers of a nautilus, or of changes in a single object over time (see Fig. 13). We have previously argued that magnification is basic to our visual construction of our environment for it provides our perceptual basis for object identity as objects move closer or further away (Confrey and Smith, 1989; Smith and Confrey, 1989). Magnification often involves combining multiplicative units to obtain a new unit. Thus a microscope with 4× eyepiece and a 300× lens produces a 1200× magnification.

6. *Order of Magnitude.* Order of magnitude is commonly used in scientific fields as a multiplicative unit, a multiplication by 10. For example, astronomers describe the distances between stars in terms of orders of magnitudes and biologists use the same units in describing, for example, populations of microscopic organisms.

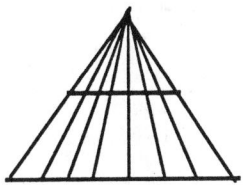

Fig. 13. Magnification.

For example a plant pathologist described comparisons among his laboratory bacteria cultures in terms of orders of magnitude, describing orders of magnitude as a primary indicator of differing population levels (Alan Maloney, personal communication). For him an order of magnitude is an appropriate segmentable unit for describing change. Other commonly used units based on multiplications by ten include the Richter and pH scales.

7. *Probability.* Probability can be viewed as a multiplicative unit in the sense that it provides a repeatable expectability between two events. Just as the constant probability that the nuclei of any atom will decay gives rise to our exponential models of radioactive decay, the expected return of anyone going to Las Vegas with a fixed sum of money who plays repeatedly in the same game exponentially approaches zero.

8. *Half-life and doubles.* Half-life is common in descriptions of residuals of a depleting quantity over time. For example, we talk about the half-life of a substance undergoing radioactive decay or the half-life of a toxic substance in the environment. In scientific contexts, doubling is often used to the change in a population. Research has shown that children from an early age use doubling as a description of change, and as a way of creating 'coordinated counts' (Confrey, 1993).

We do not claim that this list is complete or that the examples are necessarily exclusive. We have presented it for two reasons: first to describe the types of situations in which we see multiplicative units as relevant, and second to begin to develop a language to talk about the actions we take in creating multiplicative units.

Issues of Commensurability of Conversions of Units

Returning to the example of growing cells, we could imagine that the biologist might choose to focus on the issue of doubling, that is, her intent might be to determine when the amount (or number) of bacteria will double (see Fig. 14). If so, we could imagine that she would make a table such as the one shown in Fig. 15. However, if she were interested in using doubles (2-splits) as a unit-name for describing her hourly unit (a 9-split), she would have more difficulty. As

d	c
number of doubles	number of cells
0.00	9.00
1.00	18.00
2.00	36.00
3.00	72.00
4.00	144.00
5.00	288.00

Fig. 14. Doubling units.

t	Δt	c	⊗c
time	unit	cells	unit
0.0		9.0	
1.0	> 1.0	81.0	> 9.0
2.0	> 1.0	729.0	> 9.0
3.0	> 1.0	6561.0	> 9.0
4.0	> 1.0	59049.0	> 9.0
5.0	> 1.0	531441.0	> 9.0
6.0	> 1.0	4782969.0	> 9.0

Fig. 15. Covariation and rate.

previously mentioned, using a 3-split as a unit name allowed her to think of a 9-split as two 3-splits. In this sense a 3-split is *one half* of a 9-split. The question now before her is how many doubles make a 9-split. From the table in Fig. 15, one can see that it will be between 3 and 4 (the first 9-split makes a total of 81 cells). However, there is no rational solution to this problem, for the number of 2-splits needed to make a 9-split is irrational (and could be calculated as $\log_2(9)$). In line with Nicole Oresme who discussed this issue in the 14th century (Smith and Confrey, 1989, in press; Oresme, 1966), we will describe two multiplicative units as *commensurable* if one unit can be created by repeating the second unit a rational number of times.[7] We will call two multiplicative units incommensurable when this is not the case. In this sense, $\sqrt{3}$ and 3 are commensurable multiplicative units, for repeating $\sqrt{3}$ two times creates the 3-unit. However 2 and 3 are incommensurable for neither can be created from the other through a rationally iterative process. Because of the incommensurability of common multiplicative units, particular attention needs to be paid to the issue of unit conversion when working with students in multiplicative settings.

RATE OF CHANGE REVISITED

Having established the viability of a multiplicative unit, we can now return to our discussion of rate of change. In the exponential example above, we identified three meanings of the term rate. Are these all legitimately called rate? To answer this,

we propose the following approach to rate: *A rate is a unit per unit comparison.*
The focus in this definition is on the word 'per'. By it we mean:

1. At a primitive level, 'per' means one unit for one unit, 'one of this for one of that'. This is an extension of a thesis by Schorn (1989) who documented young children's command of a closely related idea, 'so much of this for so much of that', however, we insist that the child recognize each block as a composite unit.

2. This primitive understanding of 'per' can be seen as evolving from building down two columns of a table, a 'covariational approach'. We believe that the covariation approach is central to the rate concept. One recognizes that constructing rate involves recognizing the repeated action in each column (+1 in the t-column; *9 in the c-column) as a 'unit', understanding that one wants to compare these units as one ℝ-unit of cells for one Δ-unit of time, and seeing that as one moves down the *unit* columns (Δt and ℝc) one is always comparing one unit to one unit.[8]

3. One builds on this primitive by combining units. The 'per' relationship is maintained whenever an action which joins units together in one unit-column is simultaneously mirrored joining units in the other unit-column. For example three Δ-units of time correspond with three ℝ-units of cells – or a change in time of 'plus 3' corresponds with a change in cells of 'times 729'. One can combine units through a variety of actions, for example *splitting* (combining units by doubling, tripling, halving, etc.), squaring, etc. This leads initially to the certainty that any n:n relation of units maintains equivalent rates where n is a positive integer.

4. These combining actions can, in general, be extended to create combinations of rational number of units, extending the equivalence understanding to p/q: p/q combinations, provided one can build fractional units. In the example above, we will be able to describe a comparison of $\frac{1}{2}$Δ-unit to $\frac{1}{2}$ℝ-unit and see it as equivalent to a unit per unit comparison provided we have an action that can create half a unit in both dimensions.

This definition of rate has similarities and differences with the treatment of rate within the research community. Kaput *et al.* (1986), Behr *et al.* (1983), and Schwartz (1988) have described rate as a quantity per one unit comparison, an intensive quantity. We have modified this description in two ways: Firstly, their unit reference seems to be focused on standard additive units for both quantities and lacks a discussion of the construction of that unit. As a result, the multiplicative rate is eliminated from the frame of reference. Secondly, for us, the numerator of the rate is no less a unit than the denominator in that it too expresses a repeated action. We focus on the mental construction of the unit, rather than on what is typically known as 'units analysis' involving standard units.

Although Thompson (1990, in press) describes a ratio as a comparison of two quantities multiplicatively, his description seems to eliminate or neglect the possibility for rates composed of multiplicative (or non-additive) units. Thus he claims that: "a specific conceived rate is... a linear function that can be instantiated with the value of an appropriately conceived structure" (1990, p. 13) and restricts his discussion of rate to additive situations (speed as distance per unit time). Although he stresses that we can create multiplicative comparisons and multiplicative combinations of quantities, his quantities themselves are never built of multiplicative units. We do not necessarily believe that Thompson himself excludes the possibility of multiplicative units, but do suspect that his use of the term 'quantity' has an unintentional outcome; readers infer that a quantity is a kind of difference (typically a difference from zero) and is, thereby additive.[9] We believe that an understanding of multiple ways of constructing units provides the basis for rate as a unit per unit comparison by establishing the role of units in both parts of the 'per' comparison.

Kaput and West (in press) differentiate between a particular ratio and a rate-ratio. They suggest that an "ingredient of a genuine rate conception is understanding of homogeneity" (p. 5). Homogeneity, they argue, is "the idea that all purchases [for example] are governed by the same price, no matter how that price may be stated, or all samples of homogeneous mixture have the same ratio of ingredients. Equivalence, numerical and semantic, applies only when the property of the situation being modeled is homogeneous in that situation" (p. 5). Homogeneity serves the purpose of establishing the invariance "in the situation and not the description" and securing numeric and semantic equivalence.[10]

We suggest that the covariation approach is a central part of how we construct homogeneity in rate. For example, if one is traveling at 30 miles per hour and asks what happens over a four-hour period, we typically will understand that we have four times as many hour-units and thus must also have four times as many distance units. Since four 30-units is 120, we state that 120 miles per four hours is the same as 30 miles per hour. Thus the unit per unit rate construct does provide for the homogeneity which Kaput and West emphasize as central to the rate construct.

However, we do differ from both Kaput and West and from Thompson in the distinction between ratio and rate. A ratio for them is described as a particular example of a rate-ratio relation. A rate concept requires homogeneity, shown as the invariance across the apparent variability as illustrated across the top of the table in Fig. 16. A ratio is any particular entry in a given cell interpreted within a situation. A student who thinks a mixture selling for eight dollars per pound can only be bought by the single pound would fail to have a rate concept, but would have a ratio concept for these researchers.

For us, the *ratio* concept is defined as the invariance across a set of equivalent proportions, a position we take in concert with the Greeks. In an earlier paper, we documented how through a process called anthyphairesis the Greeks created a representational symbol system in which the notation for any pair of segments in a given ratio, a:b would be notated as 'the same' (Smith and Confrey, 1989). Thus for us, the homogeneity concept (going across the table in Fig. 16) is essential to *both* rate and ratio. However, what is central to an understanding of rate for us is the sense that it can vary (moving vertically in Fig. 16) as a quantity itself.

We believe that the ability to recognize *variation* in a rate of change is essential for the transition to calculus, and is the point where ratio and rate concepts depart ways. In calculus, all non-linear functions exhibit varying rates of change.[11]

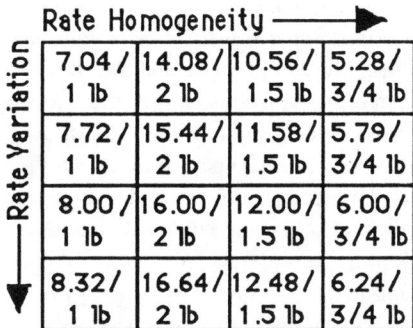

Fig. 16.   Price rates.

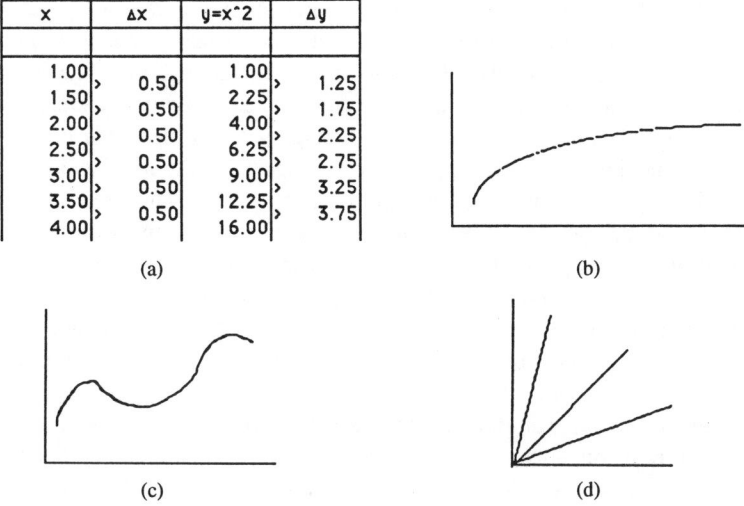

Fig. 17.   Variable rates: (a) increasing, (b) decreasing, (c) varying, and (d) varying constant.

We are concerned that placing the emphasis in defining rate as a description of the invariance across equal proportions may create an obstacle to understanding varying rates of change for it tends to neglect the importance of one's experiential understanding that a rate can vary over time (for example, an understanding of the changing speed for a freely falling body). We seek to establish a rate construct that allows one to explain both the uniformity of unit to unit comparisons (homogeneity) and the variation in rates over time (non-homogeneity). This is essential to us because our work has concentrated on the development of rate within a variety of families of functions, most notably, linear, quadratic and exponential. Within these environments, one must explain rates which vary, such as those Fig. 17.

For example, we find students examining differences in the y-values of a quadratic function, and, in comparison to a linear function, describing it as an increasing rate (Fig. 17a). We would argue that the students do not see a homogeneity in the situation in terms of a constancy in the rate, yet use the word 'rate' in the situation appropriately to imply a description of a potential repeatability. Thus the unit 1.25 (in the Δy column) per unit 0.50 (in the Δx column) is conceived of as repeatable. The recognition that it does not repeat, that this unit per unit is not large enough to describe the next unit change leads them to describe the situation as having an increasing rate. We have not encountered any student spontaneously describing this as having an increasing ratio, suggesting that ratio and rate may mean different things in this context. In the other figures, students can explain that the rate of the graph is decreasing (Fig. 17b), or the rate varies (Fig 17c). In Fig. 17d, one can describe it as having either an increasing rate or increasing ratio as one moves from the lowest spoke to the highest one. We suggest that the variation across rates of change should be introduced earlier in order to establish a more robust rate construct. We further believe that they pose a significant epistemological challenge to our understanding of the rate construct.

It appears that there is evidence of a primitive, experiential understanding of varying rates of change. Children can distinguish more, less and perhaps the same (to some degree of accuracy) rates of change, even though they may not be able to discuss them in terms of units, or of unit per unit. Thus, they possess an experiential dimension of rate as extensive, but need not have made the connection to the ratio/rate, unit per unit, approach. This experiential dimension allows them to compare rates non-numerically as more, less or the same, which is essential to experiencing something as a rate at all. A child recognizes a change of speed in an automobile; a gust of wind is heard as 'blowing harder', both implicit rate concepts.

Volume turned up, running hard to end a race, breathing slowing down after a rest, are all rate concepts to children. Discrete cases also are common, although their comparisons may be more difficult for children to articulate. When it rains or snows harder, they perceive more flakes falling in a given visual frame; they prefer the cookie brand known to have more chips per cookie, etc. Following the distinctions of Davydov (in Steffe, 1991) we use the term 'comparative'; thus, we refer to this as a 'comparative dimension'.

One may question whether such an understanding is pre-mathematical or mathematical. Basic to our argument for the importance of multiple representations is the claim that it is through the ability to coordinate representations that one engages in mathematics. Graphs provide an environment in which children demonstrate their ability to treat this experiential dimension as mathematically situated. Tierney and Nemirovsky (1991) have demonstrated elementary students' competence in displaying and justifying a series of changes graphically (the addition and subtraction of blocks into and out of a paper bag, the accumulation of rain water and the description of plant growth).

Furthermore, in work with secondary students, Nemirovsky (1991) has suggested that the transition to calculus could be facilitated if students were to become facile at identifying and working with six prototypic shapes of curves, each demonstrating a different pattern of variation in rate of change. These analyses suggest that a second understanding of varying rate lies in the coordination of one's experiential knowledge of slope and changing slope (as steepness) with its visualization on a graph. We suspect that students form a direct connection between slope and rate of change which is not mediated by numeric analysis. This second rate concept is more holistic than the analytic 'unit per unit', and connects the experiential basis of slope with rate variations in contextual problems.

An interesting question is how to extend the kind of careful analysis done by Thompson and Kaput and West to describe/produce an analysis of unit per unit change in a situation in which the rate is experienced as variable, either as a comparative dimension or visually through slopes of graphs.[12] This returns us to a consideration of varying rate in the covariation, tabular approach discussed earlier.

In extending the unit per unit analysis to varying rates, one can consider how a rate structure, or price, of a mixture can be varied by increasing/decreasing the unit in one column (while holding the other unit fixed), or vice-versa. These are analytic/numeric strategies, and our students find them intuitively easy to express within the table window on Function Probe *if a constant unit change* can be created in one of the two variables (x or y). We have found that students can frequently visually link these curves with their descriptions of situations and discuss rate in this context. For example, Rizzuti (1991) writes of "how several students spontaneously generated the construct of rate of change of a function (both additive and multiplicative) in their explanations of why graphs behaved as they did" (p. 123). She cites an example where a student predicted the shape of a geometric sequence:

All right, this would be linear, because you're multiplying by the same – No, it would not be. No, take that back. Because 512 to 3072 is smaller than from 3072 to 18000 so you're gonna have this slope again. It's gonna go to a real steep slope. [p. 126]

and later,

I believe that we get the curve on the other one because its multiplication and when you multiply numbers they get bigger faster. And then when you add numbers or subtract numbers its always constant. It's always like minus 103, so the slope is actually just a little bit smaller. The slope is smaller here than it is here. I mean, it starts out small, the slope on the first one [the geometric sequence] and then it goes real big. And then the slope on the right one here [the arithmetic sequence] – the straight line – it just remains constant the whole time. [p. 129]

In this example, one sees the student appealing to both a unit per unit comparison and then a visual perception of slope.

53

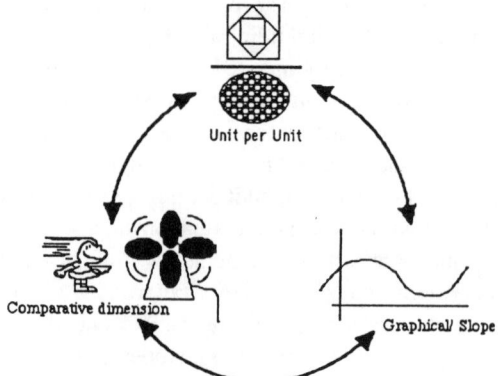

Fig. 18. Rate as a coordination of multiple concepts.

An Integrated Understanding of Rate

We are suggesting that the kind of rich understanding of rate that will serve students well, particularly as they enter calculus, involves all three components we have discussed (Fig. 18). We argue that:

1. The unit per unit comparison is central to a construction of rate and that understanding a unit as an intentional segmenting of experience in a potentially repeatable manner is a necessary first step in making this comparison. Thus, we express the centrality of homogeneity to rate, particularly in situations where rate is constant (additive *or* multiplicative contexts), but through a units-based approach.
2. Children have a 'comparative dimension' for varying rates, which allows them to assert greater, lesser and constancy of rate.
3. Graphs which allow for an experiential interpretation of slope provide an alternative expression for varying rates at an early age.
4. A coordination of these multiple representations of rate will be necessary for a more robust concept of rate.

Rates of Change in Exponentials Revisited

Early in the paper, we described three separate views of rate within the biological cells example. We can now revisit them using the theoretical approach to units and rates.

1. *An additive rate of change.* Looking at the exponential function from the perspective of an additive rate of change, one concludes that the rate is varying. In fact, for increasing exponentials, the drama of the increasing magnitude of the rate of change is an essential component to being mathematically literate about the exponential. Several scientists have described their qualitative understanding of an exponential function as involving a situation in which a population decreases

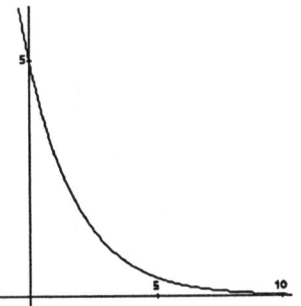

Fig. 19. Approaching an asymptote.

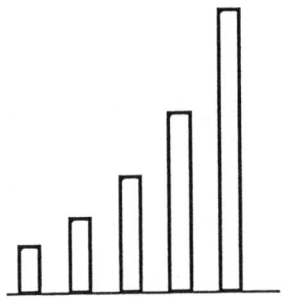

Fig. 20. Multiplicortive bars.

rapidly at first, then tends to level off and asymptotically approach a lower limit (Fig. 19). Empirical evidence has confirmed that students can, in fact, coordinate their recognition of an increasing (or decreasing) additive unit and the curvature of a graph placed on an additive scale.

2. *A multiplicative rate of change.* The analysis has legitimized a multiplicative rate of change approach by recognizing the multiplicative unit. Visually, on an additive scale, it is difficult to see the exponential curve's defining quality; although Carlstrom (1992) has demonstrated that students are quick to recognize a visual regularity when working in a computer environment in which exponentially growing bars are generated (Fig. 20). Building concepts of multiplicative rates constructed from multiplicative units should play a central role as students work on understanding how multiplicative worlds generate constant doubling times and constant half-lives.

3. Although the 'proportional new-to-old' description of rate of change described previously seems, in a way, to be neither here nor there, that is it does not quite fit into either an additive or a multiplicative understanding of change, we believe that it is key in assisting students' move into calculus. In Fig. 21, we have redrawn

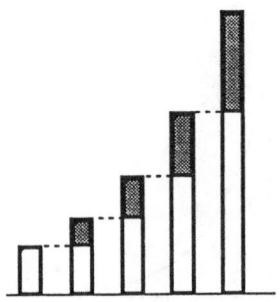

Fig. 21.   Proportional new-to-old.

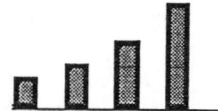

Fig. 22.   Proportional new.

Fig. 20 to illustrate the proportional new-to-old rate. As one moves from one bar to the next, what is new, is the shaded area of the next bar. This also represents the additive change (difference) from bar to bar. However, one can also see in Fig. 22 that the additive increments are also increasing. As one investigates further and finds that these added increments are increasing by the same constant ratio as the original bars (also see Fig. 3b), it is only a short step to argue that the additive change (the shaded areas) will always be the same proportion of the whole bar. This principle becomes a fundamental property of exponential functions in calculus, that the derivative of any exponential function is directly proportional to the function itself. In formal terms, this can be expressed as:

1.   If $f(x) = a^x$, then $f'(x) = k * a^x$, where k is constant for fixed a.

We would suggest that building this understanding should precede the introduction of e, for as we have already discussed, e cannot be constructed as a segmentable multiplicative unit. Thus, we expect e to be constructed out of its relationship to equation 1, since it has the unique property that when a=e, then $k = 1$ or:

2.   If $f(x) = e^x$, then $f'(x) = 1 * e^x$.

Within this setting, multiple approaches can be developed for constructing e and the natural log function. We plan to explore this issue further in a future paper.

CONCLUSIONS AND FUTURE DIRECTIONS

In this paper we have presented an epistemological discussion for understanding rate and unit that builds on multiple approaches. The description of a multiplicative unit is based on language and actions we have observed in our work with students and scientists. We believe this description has a potential to bring a rich set of concepts and language into our mathematical discussions that do not fit into standard additive/quantitative models. Likewise, although we do not have serious disagreements with others who have described rate concepts, we believe that the tendency to limit rate to situations comparing quantity to (additive) units unnecessary restricts rate and does not capture the variety of situations and mental actions which people need and use in building a rate concept. Furthermore, we argue that an understanding of varying rates of change must accompany the development of the homogeneity of the rate concept. The multi-conceptual approach to rate which we have proposed holds particular promise in preparing students for calculus.

ACKNOWLEDGEMENTS

We would like to acknowledge that this project was funded under grants from The National Science Foundation (MDR-9053590) and Apple Classrooms of TomorrowSM, Advanced Technology Group, Apple Computer, Inc.

NOTES

[1] The potential role of technology in our changing concepts of function cannot be ignored. For example, with Function Probe© (Confrey, 1991) a student can use the method of co-filling two columns to generate a functional relationship. The ready access that the software provides for students to create and explore various patterns in tables broadens the opportunities for them to develop a rich covariational concept of function that would be much more time-consuming and difficult with paper and pencil.

[2] This is not intended to be an exhaustive description of concepts of rate of change. In particular the r in equations such as $P_f = P_i e^{rt}$, often used in population, radioactive decay and interest problems encourages us to explore additional concepts of rate-of-change. We do not have the space to discuss this further in this paper, but will address this issue and the development of e in a follow-up paper.

[3] Many people suggest that this is division and not multiplication. We are arguing that the operations of multiplication and division are paired, and that splitting is a description of the overriding primitive that can be used in both cases. However, we identify the units as multiplicative units in the general case.

[4] Although we describe the outcome in terms of number of 'cells', it is used as a way of describing what one gets after the split in a language that allows for a single naming system.

[5] Dienes (1967) makes the distinction between a state and an operator. We operate to create states. State language is easy to use to describe the outcome of a unitizing action. However, as already discussed, creating the unit as 'state' simultaneously involves creating the unit as the repeated action, the successor operation.

[6] We refer to this world of activities as 'splitting world'. We use this overriding description because we like to contrast the primitive 'split' with the primitive 'count'. However, we will also use split as a particular kind of multiplicative unit (see #1).

[7] Commensurable magnitude was used by the early Greeks to mean 'have a common measure', a common unit that could be used to evenly measure each magnitude. The issue of incommensurability first came up for the Greeks in the investigation of the diagonal and side of the square by the Pythagoreans (Smith and Confrey, 1989).

[8] We will use $\triangle$-unit and Ⓡ-unit to denote additive and multiplicative units respectively.

[9] Carlstrom (1992) has suggested that a way to avoid the additive inference about quantity is to switch to using the term dimension where a dimension is constructed as the progressive integration of units where the units can be constructed using the multitude of ways discussed in this paper.

[10] A difficulty we have with Kaput and West's description is their attribution of homogeneity to the situation, leaving out the active and intentional construction of the situation by the individual. In this sense, we agree with Thompson when he states: "When we shift our focus to the operations by which people constitute 'rate' and 'ratio' situations, it becomes clear that situations are neither one nor the other. Instead how one might classify a situation depends upon the operations by which one comprehends it" (in press, p. 11–12). Although this may be what Kaput and West are suggesting by the idea of semantic equivalence, we believe that a semantic analysis is insufficient if it does not explicitly allow the situation to be viewed in terms of the actions and intentions of the individual.

[11] This is partly because in conventional calculus instruction, rate of change is built *only* on additive units. Whereas examining an exponential rate as a multiplicative unit per additive unit leads to a constant rate of change, analyzing an exponential rate as an additive unit per additive unit leads to a varying rate of change.

[12] A transition towards instantaneous velocity and the derivative is implied as a tangent or as a algebraic derivative is implied.

REFERENCES

Afamasaga-Fuata'i, K.: 1991, *Students' Strategies for Solving Contextual Problems on Quadratic Functions*, unpublished doctoral dissertation. Cornell University, Ithaca, NY.

Behr, M., Lesh, R., Post, T. and Silver, E.: 1983, 'Rational-number concepts', in R. Lesh and M. Landau (eds.), *Acquisition of Mathematics Concepts and Processes*, New York, pp. 91–126.

Borba, M.: 1993, *Students' Understanding of Transformations of Functions Using Multi-Representational Software*, unpublished doctoral dissertation, Cornell University, Ithaca, NY.

Borba, M. and Confrey, J.: 1992, 'Transformations of functions using multi-representational software: Visualization and discrete points', a paper presented at the *Sixteenth Annual Meeting of Psychology of Mathematics Education-NA*, Durham, p. 149.

Boyer, C. B.: 1968, *A History of Mathematics*, New York.

Carlstrom, K.: 1992, *Units, Ratios, and Dimensions: Students' Constructions of Multiplicative Worlds in a Computer Environment*, unpublished thesis, Cornell University, Ithaca, NY.

Confrey, J.: 1990, 'Splitting, similarity, and rate of change: A new approach to multiplication and exponential functions', a paper presented at the annual meeting of the *American Educational Research Association*, Boston.

Confrey, J.: 1991a, 'The concept of exponential functions: A student's perspective', in L. Steffe (ed.), *Epistemological Foundations of Mathematical Experience*, New York, pp. 124–159.

Confrey, J.: 1991b, *Function Probe©*[Computer Program], Santa Barbara.

Confrey, J.: 1991c, 'Learning to listen: A student's understanding of powers of ten', in E. von Glasersfeld (ed.), *Radical Constructivism in Mathematics Education*, Dordrecht, pp; 111–138.

Confrey, J.: 1992, 'Using computers to promote students' inventions on the function concept', in S. Malcom, L. Roberts and K. Sheingold (eds.), *This Year in School Science 1991: Technology for Teaching and Learning*, Washington, DC, pp. 141–174.

Confrey, J.: 1993, 'Learning to see children's mathematics: Crucial challenges in constructivist reform', in K. Tobin (ed.), *Constructivist Perspectives in Science and Mathematics*, Washington, DC: American Association for the Advancement of Science, pp. 299–321.

Confrey, J., in press a, 'Voice and perspective: Hearing epistemological innovation in students' words', in N. Bednarz, M. Larochelle and J. Desautels (eds.), *Revue des sciences de l'éducation*, Special Issue.

Confrey, J.: in press b, 'Splitting, similarity, and rate of change: New approaches to multiplication and exponential functions', in G. Harel and J. Confrey (eds.), *The Development of Multiplicative Reasoning in the Learning of Mathematics*, Albany.

Confrey, J. and Smith, E.: 1989, 'Alternative representations of ratio: The Greek concept of anthyphairesis and modern decimal notation', a paper presented at the *First Annual Conference of The History and Philosophy of Science in Science Teaching*, Tallahassee, pp. 71–82.

Confrey, J., Smith, E., Piliero, S. and Rizzuti, J.: 1991, 'The use of contextual problems and multi-representational software to teach the concept of functions', *Final Project Report to the National Science Foundation* (MDR-8652160) and Apple Computer, Inc.

Davis, R.: 1988, 'Is percent a number?, *Journal for Mathematical Behavior* **7**, 299–302.

Dienes, Z. P.: 1967, *Fractions: An Operational Approach*, New York.

Dubinsky, E. and Harel, G.: 1992, 'The nature of the process conception of function', in G. Harel and E. Dubinsky (eds.), *The Concept of Function* (MAA Notes V. 25), Washington, DC.

Greer, B.: 1988, 'Nonconservation of multiplication and division: Analysis of a symptom', *The Journal of Mathematical Behavior* **7**(3), 281–298.

Kaput, J., Luke, C., Poholsky, J. and Sayer, A.: 1986, 'The role of representation in reasoning with intensive quantities: Preliminary analyses', *Educational Technology Center Tech. Report 869*, Cambridge.

Kaput, J. and West, M.: in press, 'Missing value proportional reasoning problems: Factors affecting informal reasoning patterns', in G. Harel and J. Confrey (eds.), *The Development of Multiplicative Reasoning in the Learning of Mathematics*, Albany.

NCTM: 1989, *Curriculum and Evaluation Standards for School Mathematics*, Reston.

Nemirovsky, R.: 1991, 'Notes about the relationship between the history and the constructive learning of calculus', *Proceedings of the Segundo Simposia Internacional sobre Investigacion en Educacion Matematica*, Universidad Autonoma del Estado va Mexico, Cuernavaca, pp. 37–54.

Nemirovsky, R. and Rubin, A.: 1991, 'It makes sense if you think about how graphs work, but in reality ...', in F. Furinghetti (ed.), *Proceedings of the Fifteenth PME Conference*, Assisi, pp. 57–64.

Oresme, N.: 1966, 'De proportionibus proportionum', in Edward Grant (ed. & trans.), *De proportionibus proportionum & Ad pauca respicientes*, Madison.

Pothier, Y. and Sawada, D.: 1983, 'Partitioning: The emergence of rational number ideas in young children', *Journal for Research in Mathematics Education* **14**(4), 307–317.

Rizzuti, J.: 1991, *High School Students' Uses of Multiple Representations in the Conceptualization of Linear and Exponential Functions*, unpublished doctoral dissertation, Cornell University, Ithaca, NY.

Rizzuti, J. and Confrey, J.: 1988, 'A construction of the concept of exponential functions', in M. Behr, C. LaCompagne and M. Wheeler (eds.), *Proceedings of the Tenth Annual Meeting of the North American Chapter of the International Group for the Psychology of Mathematics Education*, Dekalb, pp. 259–268.

Rubin, A.: 1992, *The Patterns of Quilts*, unpublished manuscript, Technical Education Research Center, Cambridge.

Schorn, A. C.: 1989, *Proportional Reasoning by Young Children*, unpublished thesis, Cornell University, Ithaca, NY.

Schwartz, J.: 1988, 'Intensive quantity and referent transforming arithmetic operations', in M. Behr and J. Hiebert (eds.), *Number Concepts and Operations in the Middle Grades*, Reston, pp. 41–52.

Smith, E. and Confrey, J.: 1989, 'Ratio as construction: ratio and proportion in the mathematics of ancient Greece', a paper presented at the annual meeting of the *American Educational Research Association*, San Francisco.

Smith, E. and Confrey, J.: 1992, 'Using a dynamic software tool to teach transformations of functions', in L. Lum (ed.), a paper presented at the *Fifth Annual International Conference on Technology in Collegiate Mathematics*, Reading.

Smith, E. and Confrey, J.: in press, 'Multiplicative structures and the development of logarithms: What was lost by the invention of function?', in G. Harel and J. Confrey (eds.), *The Development of Multiplicative Reasoning in the Learning of Mathematics*, Albany.

Steffe, L.: 1988, 'Children's construction of number sequences and multiplying schemes', in M. Behr and J. Hiebert (eds.), *Number Concepts and Operations in the Middle Grades*, Reston, pp. 119–141.

Steffe, L.: 1991, 'The constructivist teaching experiment: Illustrations and implications', in E. von Glasersfeld (ed.), *Radical Constructivism in Mathematics Education*, Dordrecht, pp. 177–194.

Steffe, L.: in press, 'Children's multiplying and dividing schemes', in G. Harel and J. Confrey (eds.), *The Development of Multiplicative Reasoning in the Learning of Mathematics*, Albany.

Thompson, P: 1990, 'The development of the concept of speed and its relationship to concepts of rate', a paper presented at the annual meeting of the *American Educational Research Association*, Boston.

Thompson, P.: in press, 'The development of the concept of speed and its relationship to concepts of rate', in G. Harel and J. Confrey (eds.), *The Development of Multiplicative Reasoning in the*

Learning of Mathematics, Albany.

Tierney, C. and Nemirovsky, R.: 1991, 'Young children's spontaneous representations of changes in population and speed', in R. Underhill (ed.), *Proceedings of the Thirteenth Annual Meeting of the NA-PME*, Blacksburg, pp. 182–188.

Department of Education
Mathematics Education
422 Kennedy Hall
Cornell University
Ithaca, NY 14853
U.S.A.

SUSAN PIRIE AND THOMAS KIEREN

GROWTH IN MATHEMATICAL UNDERSTANDING:

HOW CAN WE CHARACTERISE IT AND HOW CAN WE REPRESENT IT?

ABSTRACT.   There has been a variety of approaches to the study of mathematical understanding, and some of these are reviewed before outlining the background to the model we are proposing for the growth of such understanding. The model is explained in detail and illustrated with reference to the concept of fractions. Key features of the model include 'don't need' boundaries, 'folding back', and the complementarities of 'acting' and 'expressing' that occur at each level of understanding. The theory is illustrated by examples of pupils' work from a variety of topics and stages. Finally one of the practical applications of the theory, mapping, is explained in some detail.

BACKGROUND TO THE THEORY

There is currently much practical interest in mathematical understanding. Curriculum reform advocates in many countries cite the need for teaching mathematics with understanding. Conference proceedings and psychological and artificial intelligence literature all exhibit interest in learning and teaching with understanding. Characterising understanding in a way which highlights its growth, and identifying pedagogical acts which sponsor it, however, represent continuing problems.

There has been a wide variety in the approaches to attempting to capture the essence of the phenomenon, and we have reviewed this in detail in Kieren and Pirie (1991) and Pirie and Kieren (1992a). Various categories of understanding, including relational and instrumental, concrete and symbolic, and intuitive and formal, have been proposed (Skemp, 1976; Herscovics and Bergeron, 1988; Schroder, 1987). Alternative views of understanding in relation to cognitive obstacles (Serpinska, 1990) or in terms of mental objects and connections among them (Ohlsson, 1988) have been proposed. Pirie (1988) has called into question the use of categories in characterising the growth of understanding as it can actually be seen by an observer. She observed understanding as a whole dynamic process and not as a single or multi-valued acquisition, nor as a linear combination of knowledge categories.

It was our wish, to better describe this growth of mathematical understanding in the children that we observed in classrooms over time, that led to the development of the ideas for our theory. It was clear to us that the children we were observing exhibited some understanding of mathematics, and so our question became: 'What *is* mathematical understanding?' Our background thinking was further stimulated by the biological theory of cognition in self-referencing systems (Maturana and Varela, 1980, 1987; Tomm, 1989). Over the past three years we have discussed

Educational Studies in Mathematics **26**: 165–190, 1994.

our developing theory in a variety of forums (see bibliography). It is a theory of the growth of mathematical understanding as a whole, dynamic, levelled but non-linear, transcendently recursive process (Kieren and Pirie, 1991). This theory attempts to elaborate in detail the constructivist definition of understanding as a continuing process of organising one's knowledge structures (von Glasersfeld, 1987).

In this paper we intend to present the theory with illustrative examples and elaborate on some of its features. We will then suggest possible practical applications for the theory, and look in detail at one of these, namely mapping the growth of a child's understanding.

A MODEL FOR THE THEORY

We first published a description of our theory in 1989. Since then the fundamental structure of the model has not changed but, if you have followed our previous work, you will see that we have altered some of our labels in response to suggestions and reactions to conference presentations (Pirie and Kieren, 1989b; Pirie and Kieren, 1990). When seeking to provide labels for new conceptions one is faced with the dilemma of either choosing existent words which one can hope already convey some of the desired meaning or creating new terminology and then attempting to invest in it associations and connotations that will carry the new ideas to the reader. Both alternatives have inherent weaknesses; words already familiar to the reader may inhibit the accretion of extra meaning and allow the criticism of ideas from an inappropriate stand point. Novel words, on the other hand, bring none of the subtle background that may be needed as a foundation for new concepts. In order, as far as possible, to avoid confusion or misunderstanding in our readers, we have picked the "labels for their key categories following a contiguity relation between the concept ... in mind and one specific of the many facets of meaning ascribed to the word in everyday use". Naturally this both assists understanding of our ideas and unfortunately also "gives rise to the illusion of an easy meta-basis for criticism... against the theory from outside" (Bauersfeld, 1988). We have, for example, used the word 'image' in the labelling of two of the levels. Since evidence at these levels is frequently based on pictorial representation we run the risk that understanding at these levels is judged to be restricted to only this mode of expression, and not seen to also encompass mental imagery. We feel, however, that the concept of mental objects is firmly enough established to be comprehended within our theory. We feel that 'image' is less open to ambiguity than, say, 'idea', which also carries a little of what we wish to describe. On the other hand, the outermost level was originally labelled 'inventing', but this gave rise to criticisms that we emphatically wished to refute. We did not wish to imply that children do not 'invent' at other levels. Indeed they do. What we want to point to is a special, new activity and we feel, with Hadamard (1945), that here "the creation of a word may be and often is a scientific fact of very great importance."

Before defining the terms in Fig. 1 or describing its key properties, we offer

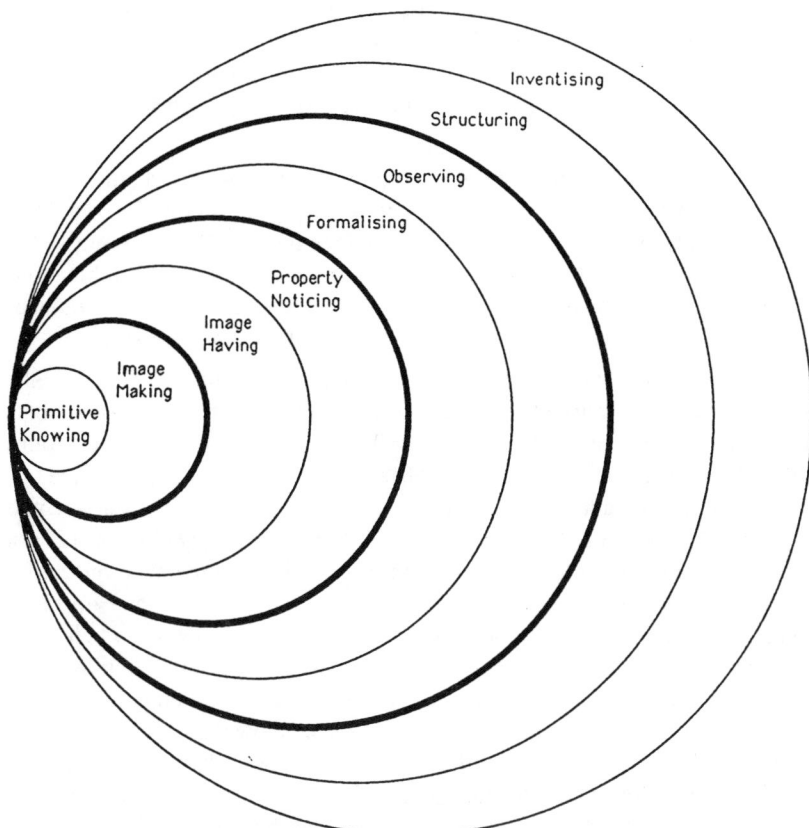

Fig. 1.

a narrative of one student's knowledge building and understanding activities as she came to understand fractions as additive quantities. (The narrative is based substantially on transcripts and student work of a class of twelve year olds. In the story of Teresa below, the work of one student in a particular fraction learning space has been extended to include possible 'observing' and 'structuring' activities.)

Teresa had been introduced in the previous school year to the term 'equivalence of fractions' and to addition of fractions, and had been given and practised using symbolic 'rules' for generating equivalent fractions and for adding them. Prior to the instruction described below, Teresa had exhibited the ability to generate fractions equivalent to 2/3, but, like most of her class, could not 'correctly' add two fractions. She said, "I think you just add the tops and the bottoms." When faced with adding more than two fractions Teresa said: "I don't know how to do that." When tested, Teresa demonstrated that she could physically make models of and identify fractions with any denominator, (1), which confirmed the teacher's assumptions that the students had usable fraction language, but did not understand the addition of fractional quantities. After the work on constructing models of fractions (through folding), Teresa's class was given a kit containing rectangles, based on a common standard sheet as a unit, representing halves, thirds, fourths, sixths, eighths, twelfths, and twenty-fourths. Teresa was given tasks such as the following.

> Using your kit notice that one fourth, three eights, and two sixteenths together exactly cover three fourths or that taken together one fourth, three eighths, and two sixteenths are equal in amount to three fourths. We can write $1/4 + 3/8 + 2/16 = 3/4$. Use your kit to find as many quantities or combinations of quantities that make exactly three fourths as you can. Draw diagrams of your findings. Write fractional number sentences like the one above.

Teresa, working with two partners, engaged in such an activity (2). Very quickly, however, Teresa internalised the activity, coming to think of fractions as numbers which described amounts (3). Teresa saw that many fractions or combinations could make up the same amount. An observer could say that Teresa's idea of addition was 'Fit the addends onto known quantities'. Thus, Teresa now could 'add' $1/3 + 1/6 + 6/12$. She thought of it as $1/3 + (1/6 + 2/12) + 4/12$. On this and many other such items Teresa identified fractional amounts or pieces on which the addends fitted or could be reconfigured to fit. It was later observed (4) that Teresa had developed a persistent and powerful strategy that when faced with complex additive (or subtractive) situations sometimes involving as many as a dozen fractional quantities, Teresa would peruse the addends looking for combinations that made up one, or one half or some other single amount.

Teresa's idea of addition was, of course, not standard and not applicable to many situations. Thus while she knew that $1/2 + 1/3 + 1/4$ was more than one, she said she couldn't "fit them" on anything. Her teacher suggested that she and some of her classmates should see if they could, given two or three fractional pieces, find one other kind of piece, replicates of which would cover all of the given pieces. At first, Teresa could not predict which piece might work, but quickly came to be reasonably skilful at it. This led to a transformation (but as seen in (4) above, not an elimination) of her previous idea of addition. Now Teresa said, "You can do $2/3 + 5/6$ because twelfths fit on both." Very quickly, Teresa, using her knowledge of equivalence, found a way of combining this new idea of addition with her knowledge of equivalence (5). When faced with the question,

> If you have an imaginary fraction kit; it has halves, fourths, fifths, tenths and twentieths, what is $1/2 + 3/4 + 2/5 + 7/10$?

Teresa says: "Twentieths will fit on all of them. Two times ten makes twenty, so one times ten or ten twentieths. Four times five makes twenty so three times five is fifteen twentieths..." Based on this kind of local, context based know-how, the majority of Teresa's classmates could by now 'add' sets of many fractions. Teresa, however, went beyond this rather concrete idea of addition, making statements like: "Addition is easy. You can make up the right kind of fractions just by multiplying the denominators and then just get the right numerators by multiplying by the right amounts. Like if you had sixths and thirds and sevenths, thirds and

$\frac{1}{2}$	$\frac{1}{3}$	$\frac{1}{6}$	$\frac{1}{12}$	$\frac{1}{24}$
1	0	1	0	0
1	0	0	2	0
1	0	0	1	2
1	0	0	0	4
0	2	0	0	0
0	1	2	0	0
0	1	1	2	0
0	1	1	1	2
0	1	1	0	4
.	.	.	.	.
.	.	.	.	.
.	.	.	.	.

Fig. 2.

sixths go together and then forty seconds work for all because sixths times sevenths give forty seconds. That will be the denominator" (6). Notice that this method is not based on particular pieces but is a method which applies independently of her previous actions. Notice also that Teresa intends that this method work for "all" fractional numbers.

When faced with a situation which involved subtraction, Teresa easily developed strategies to accomplish such tasks varying from using her concrete addition strategy subtractively to making up a method for subtraction (7).

With her classmates, Teresa then worked on the task:

Using halves, thirds, sixths, twelfths and twenty fourths, make two thirds in as many ways as you can. You can use this chart to help you keep track.

At first Teresa got out her kit and started covering pieces, but she quickly abandoned the kit and started to systematically fill in the chart (8) (see Fig. 2). Once again she had made up and used a method which made no reference to actions; it just followed a symbolic pattern.

At this point, Teresa declared that there should be an exact predictable number of combinations for two thirds or indeed for any fraction (9). She now tried one sixth, one third, and one and then made up a formula which predicted the number of combinations of the fraction set which would make up a given fraction. She tested this by making charts for 4/3, 5/6, and 5/3. She said, "I bet I can predict it for one half, too." Later, Teresa and two colleagues worked on seeing how one could make up and verify general patterns which would relate a given quantity to a fraction set and combinations of that set to the quantity (10). She was on her way to working on partition theory.

We will now use the above narrative as illustration, while we define elements of our theory and describe some properties of it. There are eight potential levels or distinct modes within the growth of understanding for a specific person, on any specific topic and we will illustrate them with reference to Teresa's growth in understanding of additions of fractions.

The process of coming to understand starts at a level we call *primitive knowing*. Primitive here does not imply low level mathematics, but is rather the starting place for the growth of any particular mathematical understanding. It is what the observer, the teacher or researcher assumes the person doing the understanding can do initially. For the growth of initial understanding of addition of fractions, the teacher wished to assume that the students already knew the language and construction of individual fractions. In the story above at (1) he was testing and probing his assumptions about Teresa's basic fraction knowing and capability. Of course, one cannot ever know what this primitive knowledge is in full. From Teresa's point of view it was at least her usable knowledge of fraction words, equivalence, and part-part-whole reasoning.

At the second level, the learner is asked to make distinctions in previous knowing and use it in new ways. In the narrative above, Teresa at (2) used previous part-part-whole knowing to combine fractional quantities into other such quantities. It was the purpose of this activity to occasion Teresa's using of fractions in an additive manner and to record and reflect on those actions. We call this mode of understanding *image making*.

At (3) above we find that Teresa can act additively with fractions without having to act on the objects. We call this activity *image having*. Notice that the original probing by the teacher into Teresa's initial understanding revealed that she did already have an image, albeit an erroneous one, for addition of fractions. This has now been supplanted by a new image formed as a result of the image making activities suggested by the teacher. At the level of image having a person can use a mental construct about a topic without having to do the particular activities which brought it about. Teresa was freed from the need to perform particular physical actions in order to solve fraction addition problems. She now had characterised, developed, and brought forth her sense of the meaning of addition of fractions.

A fourth level or mode of understanding occurs when one can manipulate or combine aspects of ones images to construct context specific, relevant properties. In (5) above we described Teresa using her image of addition as finding subparts which fit and her idea of equivalent fractions to generate a means of performing addition. We call such activity *property noticing*. Notice that Teresa's 'property' is closely tied to her image of fractions – each fraction in a sum is worked on, on its own and then combined. This new 'property' of addition differs from Teresa's image of addition in that Teresa has noticed how her image of addition 'works' and is able to combine aspects of it, structure it and explain this structure. Such property noticing is also evident at (4) where Teresa is seen to have developed a useful additive heuristic based on her image of fitting fractions onto known fractions.

At the following level of understanding, *formalising*, the person abstracts a method or common quality from the previous image dependent know how which characterised her noticed properties. At (6) Teresa is observed to see that addition is something that can be done using only the number concepts and symbols related to fractions. Rather than the addends being thought of in singular image related terms, and addition being carried out dependent on these terms, addition now takes on a formal mathematical character – it is a method which works for any set of fractions without reference to their more physical quantitative meaning. At this point Teresa, and anyone formalising, would be ready for, and capable of enunciating and appreciating a formal mathematical definition or algorithm – in this case for addition. This kind of understanding occurs again in (7) for subtraction and again in (8) when Teresa substitutes building a chart through patterns on addends as numbers rather than recording particular indicated sums which correspond to particular ways of quantitatively making two thirds.

A person who is formalising is also in a position to reflect on and coordinate such formal activity and express such coordinations as theorems. We call such an understanding activity *observing*. Such activity occurs at (9) where Teresa is looking for patterns in her charts or formalisms for combining fractions. Teresa's formula for predicting the number of combinations of her fraction set which would add up to a given fraction, itself an act of *observing*, can be contrasted with an inner understanding activity expressed as "I can get other combinations for two thirds by replacing any addend with equivalent pieces." This we could call a *noticed property*. Both would contrast with an even more general but more concrete *image* expressed as, "Many combinations of fraction-pieces can make two thirds."

Structuring occurs when one attempts to think about ones formal observations as a theory. This means that the person is aware of how a collection of theorems is inter-related and calls for justification or verification of statements through logical or meta-mathematical argument. At (10) for Teresa a statement about partitioning would not be about physical chunks, that would be *image making* or *property noticing*, nor about making partition charts which would be *formalising*. In *structuring* a statement about partitions is a statement about a mathematical structure independent of physical or even algorithmic actions.

As mentioned earlier, the outermost level of the eight in our model we call the level of *inventising*. Within a given topic a person at this level has a full structured understanding and may therefore be able to break away from the preconceptions which brought about this understanding and create new questions which might grow into a totally new concept. At the *structuring* level one can see the rationals as a set of numbers with the form of an ordered pair, a/b. This set of numbers is also seen to be a quotient field. One might now *inventise* by asking: 'What might numbers with the form of ordered quadruples a/b/c/d be like?' It was just such a question which stimulated Hamilton to think about and finally develop quarternions from having the *structured* understanding of the complex numbers.

It must always be remembered that our diagram, given above, is only an attempt to represent our ideas in a 2-dimensional form. It is not 'the model' itself and, indeed, has many drawbacks, although with these caveats in mind the diagram is a useful tool, as we shall show later in this paper, in the mapping of growth of understanding. We need to stress at this point that we do *not* see the

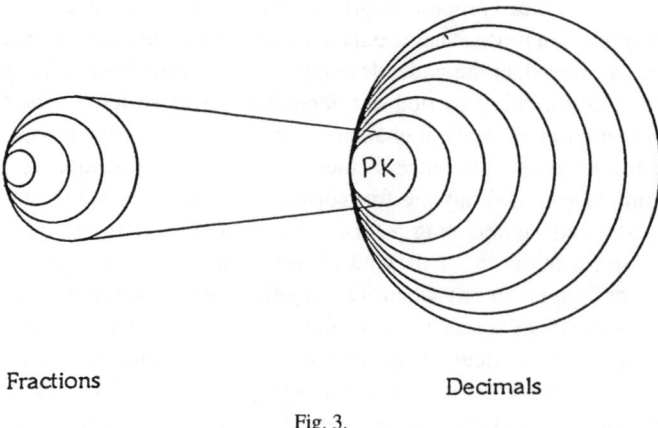

Fractions   Decimals

Fig. 3.

growth of understanding as a monodirectional process. In an effort to convey this visually, we have presented the model as a sequence of nested circles or layers, thus emphasising the fact that each layer contains all previous layers and is embedded in all succeeding layers. We see growth as represented by back and forth movement between levels and it is thus that we characterise understanding as a dynamic and organising process. We use the language of 'levels' and 'layers' and certainly there is some underlying hierarchy within the model. Just as the term primitive knowing does not imply low level mathematics, so there is no intention to link the outer levels necessarily with 'better' or 'high level' mathematics.

Earlier in this paper, we noted that primitive knowing was the background mathematical understanding needed to build an understanding of some particular concept. It is therefore possible that a full or partial understanding of that concept could then, in turn, be observed as the primitive knowing for a new mathematical exploration. The model has a fractal-like quality: inspection of any particular primitive knowing will reveal the layers of inner knowings. Using our suggested representation of the model we might illustrate a child using her current under-standing of fractions – incomplete as it is – as part of the primitive knowing for the understanding of decimals (Fig. 3).

FEATURES OF THE THEORY

'Don't need' Boundaries

One of the strengths of mathematics is the ability to operate at a symbolic level without reference to basic concepts and this is reflected in a critical element of our theory, observable in the model, and illustrated by the bold rings. Beyond these boundaries the learner is able to work with notions that are no longer obviously tied to previous forms of understanding, but these previous forms are embedded in the new level of understanding and readily accessible if needed. We call these rings the 'don't need' boundaries in order to convey the idea that beyond the boundary one does not need the specific inner understanding that gave rise to the outer knowing. One can work at a level or abstraction without the need to mentally or physically reference specific images. This does not, of course, imply that one cannot return to the specific background understanding if necessary. Indeed quite the contrary is true as will be shown in our discussion of *folding back* and disjoint understanding later in this article. We simply point to the fact that one does not need to be constantly aware of inner levels of understanding.

The first of the 'don't need' boundaries occurs between *image making* and *image having*. When a person has an image of a mathematical idea, she does not need actions or the specific instances of image making. Teresa, with a mental picture, an image of addition of fractions, stopped physically fitting and covering items with her kit. In contrast, property noticing is defined as the result of working with existent images to notice general properties and therefore access across the *image having/property noticing* boundary is essential.

The next 'don't need' boundary occurs between *property noticing* and *formalising*. A person who has a formal mathematical idea does not need an image. Teresa was able to think of fraction addition as combining entities of the form a/b with no reference to actual partitioning and covering. As with the relationship between image having and property noticing, *observing* involves, by definition, focusing on current *formalising*.

A third 'don't need' boundary occurs between *observing* and *structuring*. A person with a mathematical structure does not need the meaning brought to it by any of the inner levels. For example, Teresa would be in a position to prove theorems about addition, division, etc., of ordered pairs without any reference to what a fraction really represents.

Folding Back

The discussion so far has focused attention on the definitions of the levels and their embedded nature, and indeed these are necessary and structurally important to the theory, but a more crucial feature is that of *folding back*. This is the activity, vital to growth of understanding, which reveals the non-unidirectional nature of coming to understand mathematics. When faced with a problem or question at any level, which is not immediately solvable, one needs to *fold back* to an inner level in order to extend one's current, inadequate understanding. This returned-to, inner level activity, however, is not identical to the original inner level actions; it is now informed and shaped by outer level interests and understandings. Continuing with our metaphor of folding, we can say that one now has a 'thicker' understanding at the returned-to level. This inner level action is part of a recursive reconstruction of knowledge, necessary to further build outer level knowing. Different students will move in different ways and at different speeds through the levels, folding back again and again to enable them to build broader, but also more sophisticated or deeper understanding.

This notion fits well with our constructivist beliefs (Pirie and Kieren, 1992b),

Fig. 4.

and is best illustrated with the story of another student, Katia, ten years old and in a different class from Teresa. She has been folding rectangular pieces of paper and drawing pictures to represent cutting up pizzas (image making). From this she has formed some image for fractions (image having). Furthermore, she has noticed the property of equivalence and can construct simple fraction chains such as $1/2 = 2/4 = 4/8 = 8/16 = \ldots$ obtained from doubling (property noticing). She has also realised that like fractions can be combined, that is to say that, as a result of colouring in activities, she knows how to combine say, 3/8ths and 2/8ths to make 5/8ths. In addition she has formalised a part of her image with the statement that "writing any number over any other number will give a fraction, where the bottom number is the folded pieces and the top number is how many you have" (formalising). The question now under discussion is, 'How can one combine non-alike fractions such as halves and thirds?' The strategy of doubling does not achieve a useful equivalence which would enable fraction addition in this situation. Clearly one possible route to a solution would be for the teacher to offer the rule: 'Find a common denominator, and cross multiply to find the numerators and then add the numerators'. This would give Katia an action to perform but not necessarily any new understanding. Actually the teacher asked, "Well what *are* these things called fractions?" Katia's response was, "They came from cutting things up – usually pizzas!" and she then folded back to drawing pizzas (image making) as illustrated in Fig. 4, and re-formed an image for halves and thirds combined now with the already noticed property for creating equivalent fractions. This she did here with the explicit aim of throwing light on her newly posed problem of addition.

Once the pizzas were both divided into 1/6ths then it seemed sensible to put the 3/6ths and 4/6ths together as 7/6ths, or a whole one and an extra 1/6th. After further similar calculations she attempted to formulate an algorithm for herself and offered, "You times the bottoms and add the tops – you times the denominators and add the numberaters" (sic). At this point she was ready and able to accommodate the teacher's rule *with understanding*.

Thus from the level of formalising Katia folded back to image having to make some sense of the operation required at the formalising level. The property of equivalence was then used for the purpose of creating a meaning for addition of fractions. In fact, the original image became enriched by the idea that one can combine, as well as divide up, fractions. It would have been of no value to simply re-call previous actions. Katia needed to re-member and combine existing images to form a new way of looking. This folding back enabled a reconstruction of inner level knowing as a foundation for outer level understanding. This, and other examples will be looked at more closely later in the paper when we discuss the use of 'mappings'. A more detailed account of folding back is given in Pirie and Kieren (1991) and Kieren and Pirie (in press).

The Complementarities of Acting and Expressing

The final feature of the theory that we wish to mention here is that of the structure within the levels themselves. We believe that each level beyond primitive knowing is composed of a complementarity of *acting* and *expressing* and each of these aspects of the understading growth is necessary before moving on from any level. Furthermore growth occurs through, at least, first acting then expressing, but more often through to-and-fro movement between these complementary aspects. At any level, acting encompasses all previous understanding, providing continuity with inner levels, and expressing gives distinct substance to that particular level.

Currently, we are trying more precisely to define these complementarities at each level and intend now to present in some detail descriptions of those within the image making, image having and property noticing levels. It is important that the reader realise that we see understanding as a process and not as an acquisition or location and that the rings illustrate modes of understanding rather than outwardly monotonic phases. For this reason we have chosen the six verbs, *doing*, and *reviewing, seeing*, and *saying, predicting* and *recording*, as labels for the acting/expressing complementarities within the image making, image having and property noticing rings. The boundaries between these complementarities are represented by dotted lines in Fig. 5.

Once more we have had to choose our terminology with care and will use the medium of a classroom example to illustrate the features we are wishing to define. Nevertheless it is perhaps appropriate first to forestall certain criticism, based on misunderstanding, by elaborating a little on the terms 'acting' and 'expressing'. As will, we hope, become clear, acting can encompass mental as well as physical activities and expressing is to do with making overt to others or to oneself the nature of those activities. Although verbal expression is not strictly necessary we must always remember that it is only through such externalisation that an observer can infer the understanding that the student is constructing. Expressing is not, however, intended to be synonymous with reflecting. Reflection is frequently a component of the acting activity, since it incorporates the process of looking at *how* previous understanding was constructed. Expressing, on the other hand, entails looking at and articulating *what* was involved in the actions.

The classroom under examination this time is that of a group of 14 year olds. The general topic under consideration was quadratic equations and the particular area being explored in the sequence of lessons we are going to look at was that of the graphical representation of such equations. The teacher assumed that the students' primitive knowing would include: evaluating polynomial expressions (at least of the second degree), making tables of values, and graphing points from these tables. The initial task offered to the pupils was the following:

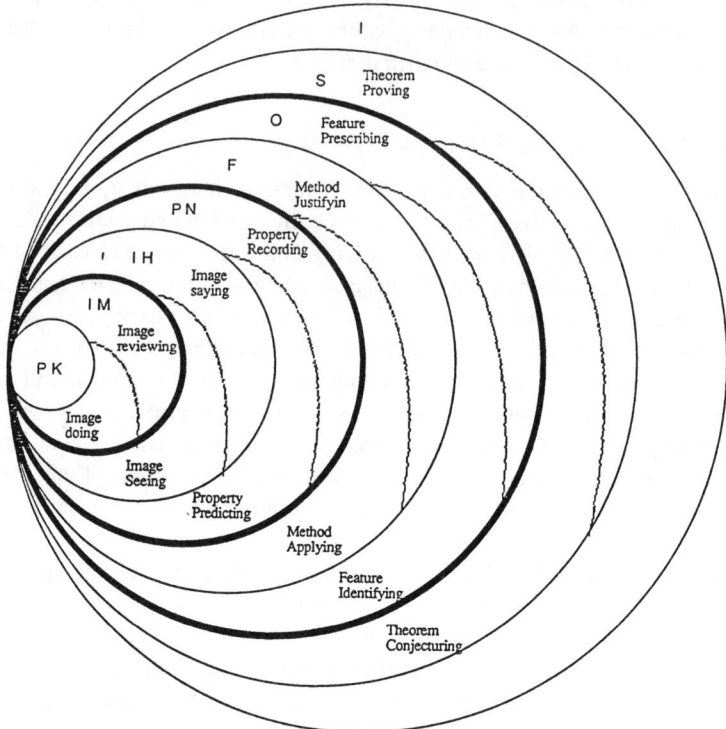

PK - Primitive Knowing
IM - Image Making
IH - Image Having
PN - Property Noticing
F - Formalising
O - Observing
S - Structuring
I - Inventising

Fig. 5.

Consider the function $y = 3x^2 + 1$.

Make a table of values for x and y, x taking values from -3 to $+3$.

Now draw the graph of the function.

Repeat this for the following functions: $y = x^2$, $y = x^2 - 2x$, $y = 2x^2 - 2x$, $y = 2x^2 - 2x - 1$.

The students were observed making tables of values, plotting and joining the points, and then moving on to the next function. Most of them successfully produced the graphs presented in Fig. 6.

Up to this point they had all been engaged in image making; more specifically in the 'acting' aspect of this level of understanding which we term '*image doing*'. They had been performing actions that might lead to the formation of an image for the graph of a quadratic function. In such behaviour one could not see whether the

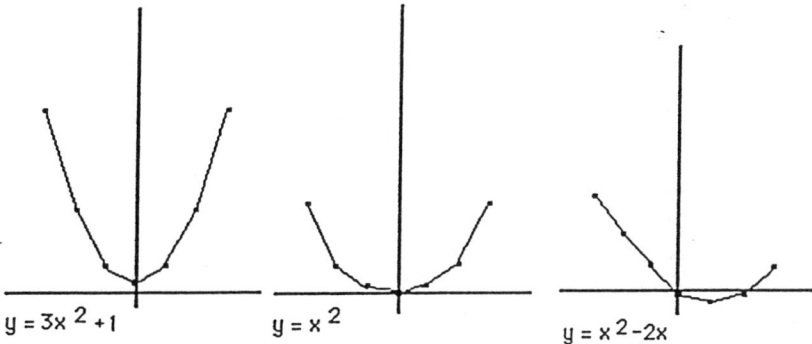

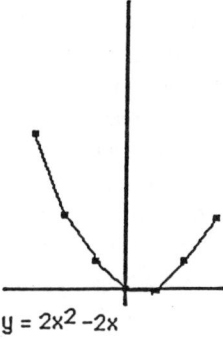

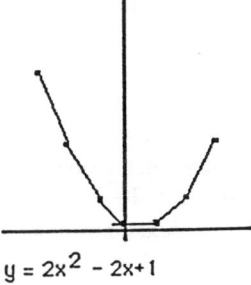

Fig. 6.

students had considered each graph as a whole before moving onto the next. This second activity is the 'expressing' complementarity at this level that we wish to term '*image reviewing*'. In this situation, 'acting' involved joining up the points in the order in which they were calculated, while 'expressing' entailed seeing some order within the activity they were engaged in. We have collected evidence that is leading to the assertion that to be said to understand a mathematical topic by showing image making behaviour, a student must have done image reviewing as well as image doing. Image doing is not enough for sustained understanding.

To probe the students' understanding, once they had had a chance to plot several of the functions given, the teacher added the point $(-2,20)$ to the first graph plotted (this point being within the range plotted (-3 to $+3$) but not on the *drawn* line) and asked the pupils to consider whether it belonged to the graph of

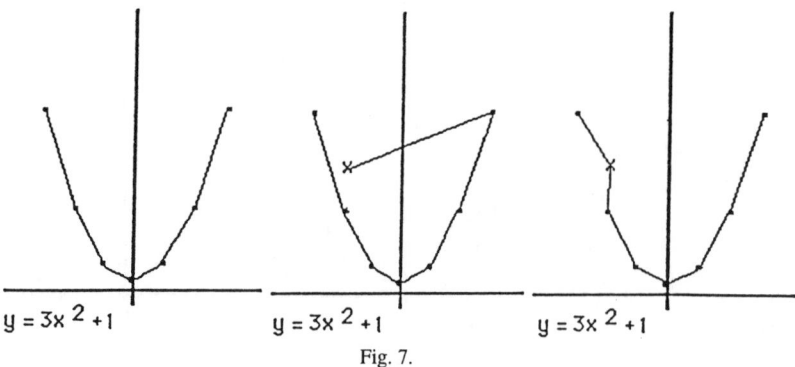

$y = 3x^2 + 1$ $y = 3x^2 + 1$ $y = 3x^2 + 1$

Fig. 7.

$y = 3x^2 + 1$ (Figs 6a,7a). The students who merely joined it to the last point plotted (Fig. 7b), were still *image doing*. They were still following instructions and had not reviewed their work, not connected their activities involving the different graphs in any way. Those students who deleted the appropriate joining line – the line joining $(-3,28)$ and $(-2,13)$ – and connected in the new point (Fig. 7c), could be seen to be engaging in *image reviewing*. They had reviewed their previous work and adapted the new task to fit some tentative idea that they might have about how these graphs should go. They were incorporating the point logically into their plotted values but not releasing it to any formed idea of the shape of a quadratic graph. What we are illustrating is that a person who is simply image doing sees her previous action as completed and rejects returning to it in anything other than a rule-bound way. The image reviewing behaviour allows for the constructive alteration of previous behaviour without yet seeing a pattern.

Those students, however, who responded with statements such as 'that can't be right', 'it can't go there', or who, when having joined up the new point, said 'that doesn't look right' were demonstrating that they had gone further in their understanding and, through reviewing the graphs they had plotted so far, had constructed some image for the graph of a quadratic. They could articulate the fact that the new point did not fit with the image they had formed, although they did not yet say why. We have called this, the 'acting' part of the level of image having, *image seeing*. The complementarity of 'expressing' at this level, *image saying*, is revealed by comments such as 'I thought they should all be U-shaped' or 'we've already got a point for $x = 2$'. The students here are able to say *why* the point does not conform to the image they have. It is interesting to note that the two student remarks given above reveal also that they have formed quite different images from the work they have been doing. The one is related to a visual representation whilst the other concerns an image of the uniqueness of points on a quadratic graph. This example serves, too, to illustrate both the fact that our use of the word 'image' is not restricted to visual images, and that for any topic there will always be a multitude of images formed. It is the interconnecting of these images that leads to the level of property noticing.

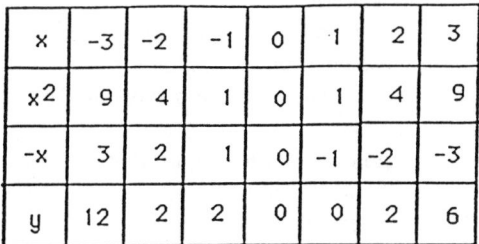

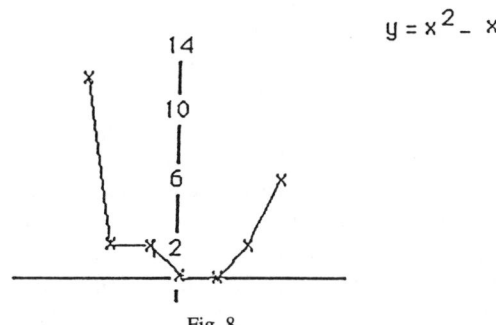

$$y = x^2 - x$$

Fig. 8.

The teacher's intervention had been deliberately intended to move the students through image reviewing to image seeing and saying, but, of course, it is the student's response to the situation that determines the effect of the questions for that person (Kieren and Pirie, 1992), and in the case of the pupils who rejected the new point or who could not express why they were not happy with the addition to the graph, the question served to confirm for the teacher that these pupils probably needed to spend more time at the image making stage. In all probability they should be encouraged to draw and review further graphs before being asked to predict features of the graph of the general quadratic function.

To clarify further the images that the students were creating, the teacher produced a table of values for the students to plot, which contained an erroneous calculation producing a point that should not be on the graph (see Fig. 8a). The student who merely plotted the points and joined them up in order (Fig. 8b) was, as above, simply image doing, following the set of instructions provided, without reviewing the outcome. Those who said "That's different" but did nothing further were image reviewing. They had made some unformulated review of their other graphs. Those, however, who responded with comments like: "I would have thought that the point should be here" or "I'll check table because it shouldn't look like that", demonstrated understanding at least at an *image seeing* level.

One student, Julie, remarked: "That is wrong. These are all pointy or flat-bottomed U-shapes", confirming for the teacher that she had moved to the level of image saying. The distinction we are trying to make between image seeing and image saying is that the former – seeing – occurs when a student has 'collected together' previous instances and has a pattern, while the image saying behaviour articulates the features of this pattern. At this level a person is in a position to talk about her actions and carry them beyond the graphing situation. The reader should notice that, even for these inner understandings, these are ways by which students can be expected to judge and defend their behaviour. They can 'organise' even these very informal schemes.

The particular example of Julie is cited to suggest that image having does not imply necessarily having the 'right' or complete picture. Two insights, however, into the images that she has, are afforded to the teacher by Julie's statement. One articulated pattern, that of the nature of the 'bottom' of the graph, is dependent on the way in which the graphs are being drawn using straight lines to join the points together. Julie is relying on her primitive knowing of graph plotting which has, to date, been always linear. The second, that of the overall shape of quadratic curves, is limited by the examples she has worked with so far. An appropriate intervention for the teacher to make at this juncture might be to ask a question to tempt Julie to fold back to further image making activity – further individual graphing. For example, 'Have you tried to graph $y = 1 - x^2$?' might be such a prompt. This new image doing and reviewing would now be informed or at least affected by the image so far seen and articulated, and would challenge the notion held as to the shape of the graph, by producing an inverted U-shape.

To this point we have tried to make distinctions between image doing and reviewing, between seeing and image saying, and between the activities of image making and image having generally. We are saying that a person showing both image doing and reviewing is showing a certain kind of understanding in their actions and we are also saying that a person who is image having is engaging in a qualitatively different kind of understanding activity, in seeing and saying that quadratics are not simply the successful results of graphing activity, but are things with identifiable features.

Returning to the example of Julie, what in reality happened was that the teacher was intrigued by the notion of 'pointy or flat bottoms' and asked her what she meant. The response, "Well it looks like a quadratic which has an odd-number in front of the of x^2 is pointy and an even-numbered one is flat-bottomed", revealed that Julie was in fact at the level of property noticing. Based on the images she had distilled from the graphs she had drawn, she had been engaging in the property noticing 'acting' activity of *property prediction*. She was distinguishing and connecting features of her image to form two classes of graphs – a new kind of understanding activity. The teacher intervention served to extend the 'acting' to the 'expressing' activity we call *property recording*. Recording here need not be written, but must involve articulate expression of some clear form. We have observed in our research several instances of where students have engaged in property predicting without recording or at least consciously making an explicit mental note that a property existed and seemed to 'work'. It seems that at both image having and property noticing levels the 'acting' notions are ephemeral and without the complementarity of 'expressing' do not remain with the student from one session to the next. A lack of 'expressing' activity seems to inhibit the students from moving beyond their previous image.

Notice again that we have selected a property – 'pointy bottoms' and 'flat bottoms' governed by the coefficient of x^2 – which is not a 'usual' property of quadratics and, in fact, may later be proven wrong or incomplete. Including under the rubric of mathematical understanding, mathematical activity which has a very non-standard character or is even 'wrong', might seem unusual, but if one considers the examples in the narrative of Teresa's work at the beginning of this essay or many other examples like it which we have gathered from children and young adults building their own mathematical knowledge structures and organising them into what, to a knowledgeable observer, would be incomplete understandings, it is apparent that such knowing and understanding is not atypical. We are looking at the nature of understanding as an activity and not as a particular content. The teacher, in a situation such as that just described, could be expected to provide the student with the opportunity to defend her 'property' by testing it against new instances which are deliberately chosen to invoke folding back, further image making, and further property noticing behaviour to allow the student to adjust and extend her image. As can be seen, our model of understanding provides teachers and researchers with a language which can enable them to look at the images which students actually 'see and say' rather than assume that students mathematical concepts correspond to given standard mathematics. We are convinced that this complementarity of acting and expressing exists and is necessary at all levels of the model and we are currently collecting data to enable us to illustrate these activities at the outer levels.

Before moving on to look at some of the applications of the theory we need briefly to return to the level of image making and counter a possible observation that this level is ill-defined since one could engage in any activity and call it image doing. We would only wish to consider potentially fruitful activity as evidence of growth of understanding. To illustrate this within the scenario of the classroom considered above, imagine the students, who, not being told explicitly to join the graph in the order of the x values, quite reasonably, from their point of view, produce something akin to that in Fig. 9. They may be engaging in a task which is congruent with their primitive knowing – from previous experience, for them 'graphs' means 'bar graphs' – but is not even image doing with respect to quadratic functions.

Applications of the Theory

With Einstein (quoted in Fine, 1986) we see a theory as: "a self-sharpening tool whose warrants and value in the end rest on this, that they permit the coordination of experience, 'with dividents' [*mit vorteil*]". So what are the 'dividends' that this theory of the growth of mathematical understanding can offer?

We have used it in a variety of learning environments as a tool to observe the mathematical behaviour of students as they work on a single mathematical task and as they build and organise mathematical knowledge structures over periods of time. The theory has enabled us to comment closely on the levels at which different students are making sense of their mathematical activities and thoughts

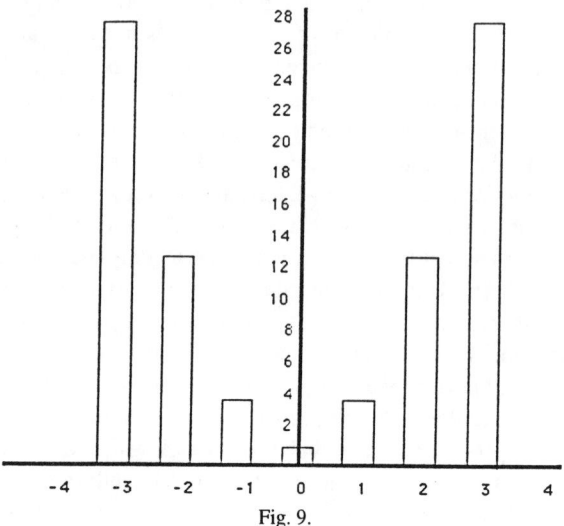

Fig. 9.

(Pirie and Kieren, 1991, 1992a; Pirie and Newman, 1990). Such insight into students' understandings have been used to provide a frame for planning and engaging in mathematics lessons and, in addition, to make observations about curriculum development (Kieren and Pirie, forthcoming).

The scope of this paper allows us to examine only one particular application of the theory in greater detail and we wish to put forward the way in which we have created a technique we call 'mapping' to record the growth of a person's mathematical understanding. Using the layered pictorial representation of the model we aim to produce in diagrammatic form a 'map' of the growth of students' understanding *as it is observed*. This last phrase, 'as it is observed', is important because we make no claims as to what might have gone on 'in the students' heads'. Analysis can only ever be based on what the teacher observes. This notion of mapping entails plotting as points on a diagram of the model, observable understanding acts and drawing continuous or disconnected lines between these points, dependent on whether or not the student's understanding is perceived to grow in a continuous, connected fashion.

To illustrate this notion, we will discuss the work of Richard, who was one of six university mathematics education students engaged for four hours in building a geometry for shapes created by a computer procedure. The students controlled the procedure by inputting three parameters from which the computer generated one member of a potentially infinite set of geometric figures such as those in Fig. 10.

Richard and his partner tried out only a few examples before he said, "Oh, they're just inward spirals." After a very few individual image making acts Richard articulated the image that he had for the geometry and we represent this in a diagram of the model by the line joining points A and B (Fig. 11). The two students then tried a few more examples to generate and test the property that

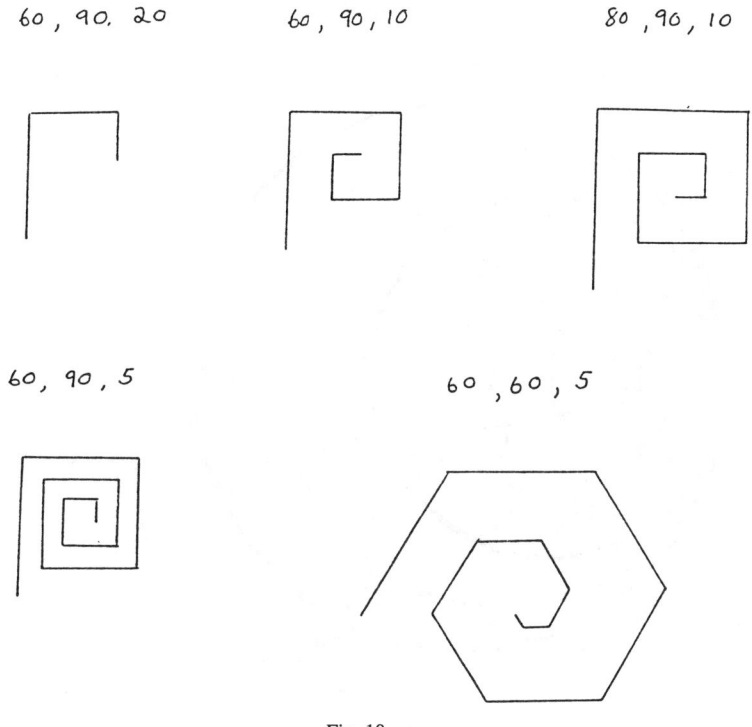

60 , 90. 20   60 , 90, 10   80 , 90 , 10

60, 90 , 5   60 , 60 , 5

Fig. 10.

an 'angle' parameter input of $360/n$ generated an n-sided polygonal spiral. We represent this property noticing with the points C and D on the diagram. While his partner now proceeded to look for other kinds of shapes. Richard stopped working on the computer. He said: "The program just generates spiral shapes by drawing a line of an input length, then turning right through the input angle. This is just repeated with the length reduced by an input decrement till it stops." This last remark suggests that he sees these shapes as a class controlled by a *formalised* statement (point E).

Richard then moved away to write-up his mathematics. He noted down the formal *observation*: "the spirals generated by the angle $180 - N$ and $180 + N$ are reflections of one another" (point F) and then set this observation in a mathematical *structure* by writing a short 'proof' based on his assumed formal procedural definition (point G). In terms of the levels of the model above, Richard was observed moving quickly and directly from *image making* through the intermediate levels out to *structuring*.

At this moment the teacher drew a square on a piece of paper and asked, "Could this be a member of your set?" To the teacher's surprise Richard said "No". The teacher then used the procedure, without allowing Richard to see the parameters that were used as input, to generate a square on the screen. Richard returned to

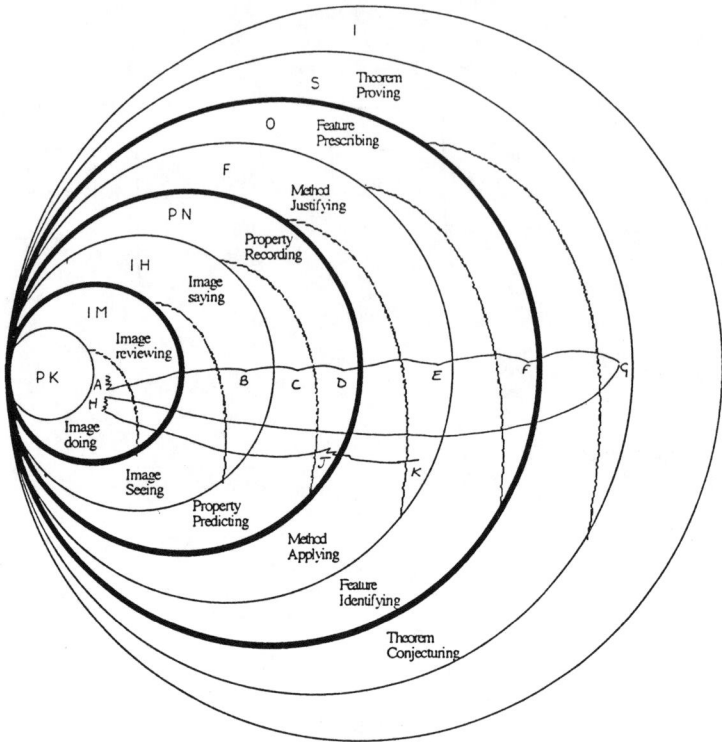

PK - Primitive Knowing
IM - Image Making
IH - Image Having
PN - Property Noticing
F - Formalising
O - Observing
S - Structuring
I - Inventising

Fig. 11.

the machine and tried out several examples before getting a square. The teacher's question had caused him to fold *back* to *image making* (point H). Eventually he *noticed a new property of* his existing image (point J) and revised his *formalisation* by saying, "Oh I see, the decrement could be zero – of course the program doesn't stop" (point K). Richard's understanding of spirolaterals had quickly grown 'deep' – out to formal, structural levels. The teacher's intervention invoked a folding back to inner level action. One might have thought that he would see the query on the square as a trivial consequence of his formalised understanding but he did not. His images at that point did not suffice to enable him to do this. He needed to reconstruct and enlarge his understanding at an inner level.

It is clear to us that student's maps are not all alike. Some students may, unlike

Richard, spend time creating a broad, rich image before moving outwards to seek properties and formalisations.

If we were to map the growth of understanding of Katia, whom we discussed earlier in this article, we would see a quite different pattern emerging. She spent several lessons in image making activities, moving forward to build up a rich image for fractions that enabled her to construct with understanding of the meaning of equivalence as it occurs in chains created by doubling the numerators and denominators of fractions. We indicate the extensive working at a single level by means of a serrated line as drawn at points A, B, and C in Fig. 12. She then folded back to further drawing and colouring-in activities (D) which led to the additional image that like fractions can be combined by counting the total number of pieces involved (E). Although not expressed in algebraic or even very mathematical language, she is heard to formalise her understanding with a generalised statement defining a fraction (F). The challenge then facing the class was to find a way to combine, or add, fractions which did not have a common denominator. Nothing in Katia's images helped her here. Had the teacher offered her the 'rule' she would have had a way of working at the formalising level, but no image in whose roots the formalising lay, to which she could fold back in later times of lack of understanding. This apparent understanding, which occurs when a student works with information that does not emerge from or become connected to her own constructured knowledge, we term *disjoint* from her existing understanding. We hypothesise, that students will be unable to successfully build further understanding based on this *disjoint* knowing until they have in fact constructed the connection for themselves. We would represent this with an unattached cross (G), to indicate that the understanding at this point was not connected to or based on the student's current understanding. In fact Katia folded back to further image making activities (H, I, J), this time with the added understanding of the images and properties she has already constructed, before moving out to the formalising level again with her personal attempt to express the process of addition (K). As stated earlier, the fact that the explanation was not completely correct does not deny the label of *formalising* for her action. When given an accurate verbal version of the process she had no trouble connecting it to her own meaning and using it with understanding.

We do not yet know whether this difference in maps is person or topic dependent. What is clear, however, is that it is not age related. One pair of students in Richard's class spent the whole of the first two hours simply making images for themselves of what the program could do. Despite repeated interventions from the teacher they resisted the need to record, or possibly even to review, their images and when they returned for the second session they were unable to recapture much of the image that they had 'seen' previously. Eventually these two students moved out to predict and record some of the properties of these images. Even with their broad understanding and multiple images of what the computer program could do, there came a point where their images were insufficient for the growth of understanding at an outer level and some folding back was necessary; while struggling to formalise some of their thinking, they constantly reviewed the

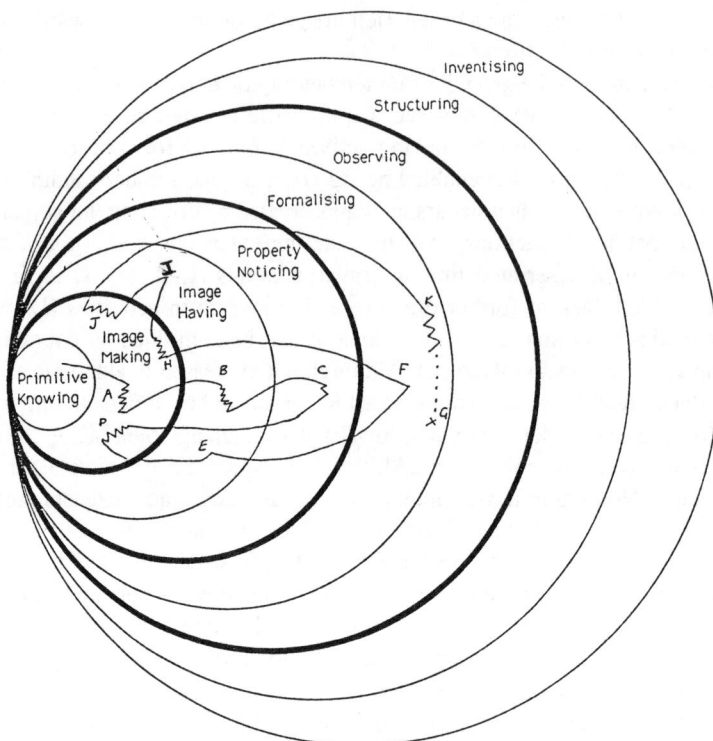

Fig. 12.

properties they had noted, for clues to a general picture and returned to drawing further spirolaterals to confirm or falsify their conjectures. The map of growth of understanding for these students would look something like Fig. 13.

The episode also illustrates the notion that folding back can happen directly to any inner level, as with Richard and Katia, or by re-tracing the path of growth through the intervening levels, as with this second pair of students. The nature of folding back cannot be *generally* prescribed; it is unique to *specific* examples of growth in understanding and to each individual person. Every student will have a singular path for any topic, and yet all paths will involve 'folding back to move out' in their actualisation.

This method of representing students' paths of growth of mathematical under-standing has the potential to allow researchers to study in detail the actual nature of this growth either for an individual over several topics, or for many students within the learning of a specified topic. The insight that this could give would be both psychologically and pedagogically valuable to the study of learning.

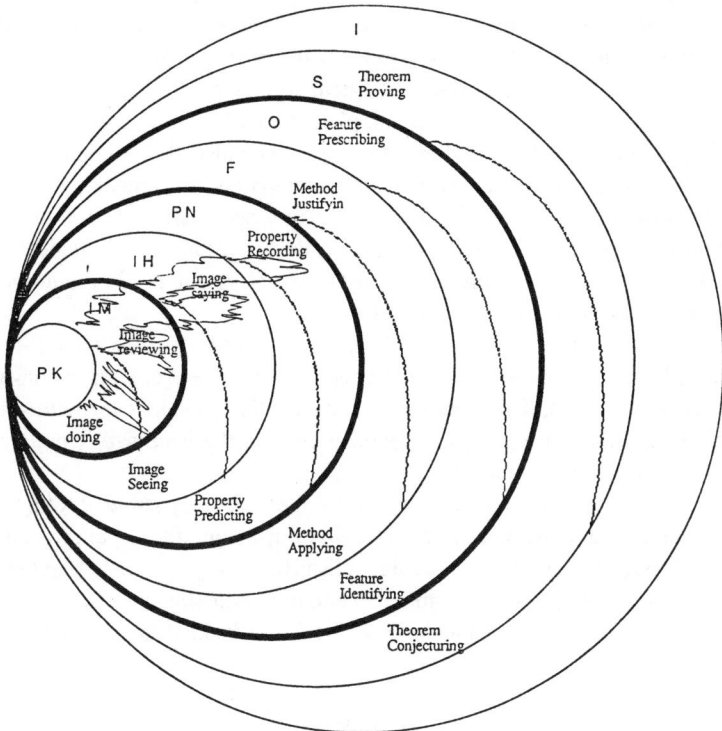

PK - Primitive Knowing
IM - Image Making
IH - Image Having
PN - Property Noticing
F - Formalising
O - Observing
S - Structuring
I - Inventising

Fig. 13.

SUMMARY

The purpose of this paper has been to show a theory of the growth of mathematical understanding which is based on the consideration of understanding as a whole, dynamic, levelled but non-linear process of growth. This theory demonstrates understanding to be a constant, consistent organisation of ones knowledge structures: a dynamic process, not an acquisition of categories of knowing.

Its levelled nature has been illustrated through the model of eight embedded rings, each of which represents a level of understanding activity potentially attainable for any particular topic by any specific person. These levels range outwards from the existing knowing that the person brings to the task, through the making and having of an image, to the noticing of properties and formalisation of that image. Reflection on the formalisation leads to observing the thought structures and to their consistent reorganisation. Logical arguments at the next level provide the necessary axiomatic structure to complete the understanding of the topic and leave the knower with the freedom to perhaps mentally alter something within that structure and explore the new field of mathematics thus created. A key tenet of the theory is that outer level knowing does not necessarily mean higher level mathematics. Equally, the converse is also true: high status mathematical topics need to be worked on at the image making level before one can begin to look for an appropriate formalisation or structure.

Other crucial features of the model include the notion of 'don't need' boundaries to explicate the unique power of mathematics to solve problems in non-image related, symbolic ways. The complementarities within each level provide the link between acting and expressing and together contain the necessary wholeness of the very nature of understanding.

This theory of growth of understanding has the built-in dynamic of folding back to move out and such growth can be thought of as a continuous path traced back and forth through the levels of knowing. This growth is a non-linear phenomenon which involves folding back to re-member and to re-construct new understanding. We see it as a non-monotonic pathway across the embedded rings in our model.

FUTURE RESEARCH

A further area for investigation that we are now examining is that prompted by many of the detailed maps of individual students that we have analysed. This is the effect of different kinds of teacher interventions on the paths of growth of understanding of their pupils. We have identified three classes of intervention, *provocative*, which have the effect of moving the student outwards, *invocative*, which have the effect of causing the student to fold back in order to enlarge or alter his image, and *validating*, which allow the teacher to view, or the student to confirm, existing understandings.

We end with a favourite quotation from Maturana:

At this point there is either much more to say ...
or nothing.

When we first enunciated our theory, we were not sure which of these options obtained. We hope now that we have demonstrated that there is indeed much more to say about the growth of mathematical understanding.

REFERENCES

Bauersfeld, H.: 1988, 'Interaction, construction, and knowledge: Alternative perspectives for mathematical education', in *Effective Mathematics Teaching*, NCTM, pp. 27–46.

Einstein, A. (as quoted in Fine, A.): 1986, *The Shakey Game – Einstein, Realism and the Quantum Theory*, Chicago, University of Chicago Press, p. 91.

Hadamard, J.: 1945, *The Psychology of Invention in the Mathematical Field*, Princeton University Press.

Herscovics, N. and Bergeron, J.: 1988, 'An extended model of understanding', in C. Lacompagne and M. Behr (eds.), *Proceedings of PME-NA 10*, Dekalb, Ill: Northern Illinois University, pp; 15–22.

Kieren, T. E.: 1990, 'Understanding for teaching for understanding', *The Alberta Journal of Educational Research* **36**(3), 191–201.

Kieren, T. E. and Pirie, S. E. B.: 1992, 'Rational and fractional numbers: From quotient fields to recursive understanding', in T. P. Carpenter and E. Fennema (eds.), *Learning, Teaching, and Asessing Rational Number Concepts: Multiple Research Perspectives*, Lawrence Erlbaum Associates, Hillsdale, N.J., pp. 49–82.

Kieren, T. E. and Pirie, S. E. B.: 1991, 'Recursion and the mathematical experience', in L. Steffe (ed.), *The Epistemology of Mathematical Experience*, Springer Verlag Psychology Series, New York, pp. 78–101.

Kieren, T. E. and Pirie, S. E. B.: 1992, 'The answer determines the question – Interventions and the growth of mathematical understanding', in *Proceedings of Sixteenth Psychology of Mathematical Education Conference* (New Hampshire), Vol. 2, pp. 1–8.

Kieren, T. E. and Pirie, S. E. B.: in press, 'Folding back: A dynamic recursive theory of mathematical understanding', in D. Sawada and M. Caley (eds.), *Recursion in Educational Enquiry*, Gordon and Bream, New York.

Maturana, H. R. and Varela, J. F.: 1980, *Autopoeisis and Cognition*, Boston University, Philosophy of Science Series, Vol. 42, D. Reidel, Dordrecht.

Maturana, H. R. and Varela, J. F.: 1987, *The Tree of Knowledge*, The New Sciences Library, Shambhala, Boston.

Ohlsson, S.: 1988, 'Mathematical meaning and applicational meaning in the semantics of fractions and related concepts', in J. Hiebert and M. Behr (eds.), *Number Concepts and Operations in the Middle Grades*, NCTM/LEA, Reston, pp. 53–91.

Pirie, S. E. B.: 1988, 'Understanding – Instrumental, relational, formal, intuitive..., How can we know?', *For the Learning of Mathematics* **8**(3), 2–6.

Pirie, S. E. B. and Kieren, T. E.: 1989a, 'Through the recursive eye: Mathematical understanding as a dynamic phenomenon', in G. Vergnaud (ed.), *Actes de la Conference Internationale, PME*, Vol. 3, Paris, pp. 119–126.

Pirie, S. E. B. and Kieren, T. E.: 1989b, 'A recursive theory of mathematical understanding', *For the Learning of Mathematics* **9**(3), 7–11.

Pirie, S. E. B. and Newman, C. F.: 1990, 'Watching understanding grow', paper presented at the Midlands Mathematics Education Seminar, University of Birmingham.

Pirie, S. E. B. and Kieren, T. E.: 1990, 'A recursive theory for mathematical understanding – some elements and implications', paper presented at AERA annual meeting, Boston.

Pirie, S. E. B. and Kieren, T. E.: 1991, 'Folding back: Dynamics in the growth of mathematical understanding', in F. Furinghetti (ed.), *Proceedings Fifteenth Psychology of Mathematics Education Conference*, Assisi.

Pirie, S. E. B. and Kieren, T. E.: 1992a, 'Watching Sandy's understanding grow', *The Journal of Mathematical Behaviour* **11**(3), 243–257.

Pirie, S. E. B. and Kieren, T. E.: 1992b, 'Creating constructivist environments and constructing creative mathematics', special edition, in E. von Glasersfeld (ed.), *Educational Studies in Mathematics* **23**(5), 505–528.

Schroder, T. L.: 1987, 'Students' understanding of mathematics: A review and synthesis of some recent research', in J. Bergeron, N. Herscovics, and C. Kieran (eds.), *Psychology of Mathematics Education XI*, PME, Montreal, Vol. 3, pp. 332–338.

Serpinska, A.: 1990, 'Some remarks on understanding in mathematics', *For the Learning of Mathematics* **10**(3), 24–36.

Skemp, R. R.: 1976, 'Relational understanding and instrumental understanding', *Mathematics Teaching* **77**, 20–26.

Skemp, R. R.: 1987, *The Psychology of Learning Mathematics* (expanded edition), Lawrence Erlbaum Associates, Hillsdale, N.J.

Tomm, K.: 1989, 'Consciousness and Intentionality in the Work of Humberto Maturana', a presentation for the Faculty of Education, University of Alberta.
von Glasersfeld, E.: 1987, 'Learning as a constructive activity', in C. Janvier (ed.), *Problems of Representation in the Learning and Teaching of Mathematics*, Lawrence Erlbaum Associates, Hillsdale, N.J., pp. 3–18.

Susan Pirie
Mathematics Education Research Centre
University of Oxford
Oxford
Great Britain OX2 6PY

Thomas Kieren
University of Alberta
Edmonton
Alberta
Canada T6G 2G5

ANNA SFARD AND LIORA LINCHEVSKI

THE GAINS AND THE PITFALLS OF REIFICATION –
THE CASE OF ALGEBRA

ABSTRACT. Algebraic symbols do not speak for themselves. What one actually sees in them depends on the requirements of the problem to which they are applied. Not less important, it depends on what one is able to perceive and prepared to notice. It is this last statement which becomes the leading theme of this article. The main focus is on the versatility and adaptability of student's algebraic knowledge.

The analysis is carried out within the framework of the theory of reification according to which there is an inherent process-object duality in the majority of mathematical concepts. It is the basic tenet of our theory that the operational (process-oriented) conception emerges first and that the mathematical objects (structural conceptions) develop afterward through reification of the processes. There is much evidence showing that reification is difficult to achieve.

The nature and the growth of algebraic thinking is first analyzed from an epistemological perspective supported by historical observations. Eventually, its development is presented as a sequence of ever more advanced transitions from operational to structural outlook. This model is subsequently applied to the individual learning. The focus is on two crucial transitions: from the purely operational algebra to the structural algebra 'of a fixed value' (of an unknown) and then from here to the functional algebra (of a variable). The special difficulties experienced by the learner at both these junctions are illustrated with much empirical data coming from a broad range of sources.

When you look at an algebraic expression such as, say, $3(x + 5) + 1$, what do you see? It depends.

In certain situations you will probably say that this is a concise description of a *computational process*. $3(x + 5) + 1$ will be seen as a sequence of instructions: Add 5 to the number at hand, multiply the result by three and add 1. In another setting you may feel differently: $3(x + 5) + 1$ represents a certain *number*. It is the product of a computation rather than the computation itself. Even if this product cannot be specified at the moment due to the fact that the component number x is unknown, it is still a number and the whole expression should be expected to behave like one. If the context changes, $3(x + 5) + 1$ may become yet another thing: a *function* – a mapping which translates every nubmer x into another. This time, the formula does not represent any fixed (even if unknown) value. Rather, it reflects a change. The things look still more complicated when a letter appears instead of one of the numerical coefficients, like in a $(x + 5) + 1$. The resulting expression may now be treated as an entire *family of functions* from R to R. Alternatively, one may claim that what hides behind the symbols is a function of two variables, from R^2 to R.

There is, of course, a much simpler way of looking at $3(x + 5) + 1$: it may be taken at its face value, as a mere *string of symbols* which represents nothing. It is an algebraic object in itself. Although semantically empty, the expression

Educational Studies in Mathematics **26**: 191–228, 1994.

may still be manipulated and combined with other expressions of the same type, according to certain well-defined rules.

At this point somebody may wonder whether the above observations on the semantics of algebra are of any practical significance. Indeed it is. For instance, when faced with such an equation as $(p+2q)x^2+x = 5x^2+(3p-q)x$ one will not be able to make a move without knowing whether the equality is supposed to be numerical or functional – whether the question is that of the value of x for which the equality holds (this value should be expressed by means of p and q), or that of the values of the parameters p ad q for which the two functions, $(p+2q)x^2 + x$ and $5x^2 + (3p+q)x$, are equal. The different interpretations will lead to different ways of tackling the problem and to different solutions: in the former case one finds the roots of the equation using the formula $x_{1,2} = (-b \pm \sqrt{\Delta})/(2a)$; in the latter, the values of parameters p and q are sought for which the coefficients of the same powers of x in both expressions are equal ($p + 2q = 5$ and $3p - q = 1$). It is worth noticing that another possible outlook – the one which leaves symbols without much meaning – may initially render the problem-solver helpless. An automatic reaction is likely to follow. A person may be tempted to do what he or she was conditioned to do when faced with a quadratic equation: regardless of the question that has been asked, he or she may have recourse to the formula for the roots.

The plurality of perspectives which one may assume while looking at such a seemingly simple thing as $3(x+5) + 1$ is certainly confusing. In the next sections it will be argued that it is also a source of algebra's strength.

1. PRELIMINARIES: ALGEBRA THROUGH THE LENSES OF THE THEORY OF REIFICATION

Algebraic symbols do not speak for themselves. What one actually sees in them depends on the requirements of the specific problem to which they are applied. Not less important, it depends on what one is *prepared* to notice and *able* to perceive. It is this last observation which will be the leading theme of the present article. The main focus will be on the versatility and adaptability of the algebraic knowledge of the student. The question that will be addressed is to what extent the learner is capable of seeing and using the variety of possible interpretations of algebraic constructs. Before tackling this problem, however, let us pause for a moment to reflect on the kind of analysis which is to be carried out here.

The distinctions which we made in the opening examples are quite subtle and refer to what is going on in people's heads rather than to what they actually communicate to the world by means of written records. Indeed, the difference between the alternative interpretations of the equation $(p + q)x^2 + 2x = 3x^2 + (p - q)x$ will not always show on a standard test, since whether the student thinks about the expression as a numerical equality or as a mere string of symbols, he or she may eventually apply the same formulae and perform the same manipulations. A painstakingly detailed scrutiny of student's behaviors and utterances (see, e.g., Schoenfeld et al., 1993; the authors call this kind of analysis 'microgenetic') is necessary to have some insight into his or her thinking.

Our 'fine-grained' analysis will be carried out within a certain theoretical framework by help of which the profusion of loose facts will hopefully turn into a meaningful, manageable whole. It will be called here *the theory of reification* (it must be pointed out that other authors may use somewhat different terminology for basically the same or closely related ideas; see the list of possible names in Harel and Kaput, 1991). Like any theoretical model, it emphasises certain aspects of the explored domain while ignoring many others. Even so, it has already proved itself as a tool for analyzing development of several mathematical concepts, notably the concept of function (Sfard, 1992: Breidenbach et al., 1992); also, it was used to introduce some order into the quickly growing bulk of findings about algebraic thinking (Kieran, 1992). Another glance at algebra through it's lenses will be offered in the next chapters.

In the remainder of this section we will confine ourselves to the presentation of the basic tenet of the theory of reification. Different elements of the resulting system of claims will be discussed throughout this paper. For a more exhaustive presentation of the basic ideas the reader is advised to turn to Sfard (1991) and Kieren (1991). Further, Dubinsky (1991) and Harel and Kaput (1991) describe closely related models. Dubinsky's ideas are an elaboration of Piagetean theory of *reflective abstraction* (Beth and Piaget, 1966); Harel and Kaput's observations bear on Greeno's notion of *conceptual entity* (Greeno, 1983). It should also be mentioned that the idea of the process-object duality of mathematical concepts which is central to this paper reminds Douady's (1985) tool-object dichotomy.

In our opening example, several mathematical *objects* have been identified as possible referents of algebraic expression. We have mentioned a number, a function, a family (set) of functions. One interpretation, however, was of a different nature: when $3(x + 5) + 1$ was read as a series of operations, it was the computational *process* rather than any abstract object (except the processed numbers) which gave meaning to the symbols. What was observed in the example apparently pervades the whole of mathematics: the same representation, the same mathematical concepts, may sometimes be interpreted as processes and at other times as objects; or, to use the language introduced elsewhere (Sfard, 1991), they may be conceived both *operationally* and *structurally*. The fact that these two ostensibly incompatible ways of seeing mathematical constructs seem to be present in any kind of mathematical activity, and thus are complementary, is the basic observation which constitutes the point of departure for our model of mathematical learning and problem-solving. The notion of complementarity is used here in the same way in which it appears (due to Niels Bohr) in physics, where to fully account for observed phenomena one must treat subatomic entities both as material particles and as waves.

The distinction between the two models of thinking, operational and structural, is delicate and not always easy to make. The ability to perceive mathematics in this dual way makes the universe of abstract ideas into the image of the material world: like in real life, the actions performed here have their 'raw materials' and their products in the form of entities that are treated as genuine, permanent objects. Unlike in real life, however, a closer look at these entities will reveal that they cannot be separated from the processes themselves as self-sustained beings. Such abstract objects like $\sqrt{-1}$, -2 or the function $3(x + 5) + 1$ are the result of a different way of looking on the procedures of extracting the square root from -1, of subtracting 2, and of mapping the real numbers onto themselves through a linear transformation, respectively. Thus, mathematical objects are an outcome of *reification* – of our mind's eye's ability to envision the result of processes as permanent entities in their own right.

This basic ontological observation has numerous theoretical implications. It generates a whole system of claims about mathematical problem-solving; it gives rise to a model of concept formation which applies to historical development as well as to individual learning; it provides its own explanation of the difficulties experienced by a student exposed to a new mathematical idea. All these topics will be discussed in what follows here, in the context of algebra. In section 2 we begin our discussion with an epistemological analysis sypported by historical observations. In section 3, we shall assume a psychological perspective and will try to show that to great extent, the model of algebra's formation constructed through the historical and epistemological analysis fits processes accompanying individual learning.

2. WHAT ALGEBRA IS AND HOW IT DEVELOPED

Many theoretical and empirical arguments may be employed to show that in mathematics, operational conception precedes the structural. What is conceived as a process at one level becomes an object at a higher level (see, e.g., Sfard, 1991, 1992). Kaput (1989) seems to make a similar observation when talking about "mental entity building through reification of actions, procedures, and concepts into phenomenological objects which can then serve as the basis for new actions, procedures, and concepts at a higher level of organization" (p. 168). Even though the point of departure for this statement may be quite different from ours (Kaput's ideas seem to bear on Piaget's theory of reflective abstraction), it points to the basic agreement about the roles of mathematical processes and objects, and about their mutual dependence. Freudenthal was one of the most outspoken proponents of the vision of mathematics as a hierarchy of alternating perspectives: "My analysis of mathematical learning process has unveiled levels in the learning process where mathematics acted out on one level becomes mathematics observed on the next" (Freudenthal, 1978, p. 33). Once again, although originating in a different kind of analysis, this assertion points to the same fundamental characteristics of mathematical construction as those implied by our theory: it underlines the fact that mathematics is a multi-level structure where basically the same ideas are viewed differently when observed from different positions.

As we have already pointed out, the process-object duality should be conveyed by algebraic constructs. For several reasons, with a common root in the rigidity of people's ontological attitudes, the ability to grasp the structural aspect is not easy to achieve. Therefore, those crucial junctions in the development of mathematics where a transition from one level to another takes place are the most problematic, and clearly the most interesting. To use Freudenthal's words once again, "If learning process is to be observed, the moments that count are its *discontinuities*, the jumps in the learning process" (p. 78). Thus, in the analysis that follows, the focus is on the singular points in the development of algebraic concepts – at those points where the ontological perspective must undergo an accommodation to make further progress possible.

Let us precede our account of the development of algebra with a remark on the nature of our investigation. The construction of algebra, like that of any other domain of human knowledge, may be scrutinized from several perspectives. One may focus on the logical structure of the discipline and ask about the way the different items of knowledge combine into one coherent system. Let us call this kind of analysis *logical*. Then, there is a *historical* approach, which concentrates on the collective efforts invested through ages into the construction of the given system of concepts. Finally, a researcher may choose to make an inquiry into the *cognitive processes* which constitute individual learning. One can hardly expect that these three kinds of analysis would yield the same result. As some writers put it, it is but a myth that "[t]he [logical] structure of mathematics accurately reflects its history" (Crowe, 1988). One should also be careful not to make automatic projections from history to psychology. After all, the deliberately guided process of reconstruction may not follow the meandering path of those who were the first travellers through an untrodden area. Even so, one may also expect some striking similarities between the pictures obtained through the different kinds of analysis. Garcia and Piaget (1989) made a particularly strong case for the analogy between the historical and psychological developments. Although much caution is advisable, logical analysis should not be dismissed altogether as a potential source of insights about the process of learning. After all, mathematics is a hierarchical structure in which some strata cannot be built before another has been completed. Thus, after making certain fundamental claims regarding the development of algebra on the basis of logical, ontological, and historical analysis, we will show in the next section that they also hold for individual learning.

2.1. *Algebra as Generalized Arithmetic: the Operational Phase*

Throughout our analysis we will call some kinds of algebra 'operational' and other kinds of algebra 'structural'. This is not to say that at any of the different stages in algebra's development only one type of ingredient – either operational or structural – was present. The complementary nature of the distinction between processes and objects makes it clear that this would be impossible. Also, it is obvious that if some processes (operational ingredient) are considered, there must be certain objects (structural ingredient) to which these processes are applied. Our claims that certain kinds of algebra were operational in their character while others were structural should be understood as referring to what constituted the *primary*

TABLE I

Rhetoric algebra – Examples

---

1. *Babylonia, second (?) millenium B.C.* (after Boyer, 1985, p. 34)

The problem: Find the side of the square if the area less the side is 14,30 (the numbers are presented on basis 60).

Solution: Take half of one, which is 0;30, and multiply 0;30 by 0;30 which is 0;15; add this to 14,30 to get 14,30;15. This is the square of 29;30. Now add 0;30 to 29;30, and the result is 30, the side of the square.

2. *Al-Khwarizmi, A.D. 825,* (after Struik, 1986, p. 58)

The problem: What is the square which combined with ten of its roots will give a sum total of 39?

Solution: ... take one-half of roots just mentioned. Now the roots in the problem before us are 10. Therefore take 5, which multiplied by itself gives 25, an amount which you add to 39, giving 64. Having taken then the square root of this which is 8, subtract from it the half of the roots, 5, leaving 3. The number three therefore represents one root of this square, which itself, of course, is 9.

---

focus of a given type of algebra. Or, to put it differently, a statement that, say, algebra was initially operational in its character means that the most advanced, central, ideas investigated at this stage were still conceived operationally rather than structurally.

The history of algebra brings much support to the thesis about the precedence of the operational over the structural approach. For thousands of years, algebra was nothing more than a science of computational procedures.

From the developmental point of view, algebra is a continuation of arithmetic. Like arithmetic, it deals (at least at its early stages) with numbers and with numerical computations, but it asks questions of a different type and treats the algorithmic manipulations in a more general way. Those who do not view symbolic representation as a necessary characteristic of algebra agree that both in history and in the process of learning mathematics, algebraic thinking appears long before any special notation is introduced. Typically, it begins with the first attempt to find the unknown number on which a given operation was performed and a given result obtained. In this kind of activity, the usual arithmetic practice of applying a computational algorithm to a concrete number must be reversed: instead of computing, say, the sum paid for a given number of pencils and a notebook, one must 'undo' what was done to the number of pencils when the total has been calculated. Although initially quite simple and intuitively immediate, such reversal stops being a trivial matter when word problems of some complexity start to appear.

Let us dwell for a moment on the earliest stages in the development of the domain. First, let us have a look at two typical pieces of ancient and medieval algebra (see Table I).

It is worth noticing that in spite of the three millennia that separate them, the two samples are not much different in their basic characteristics: they involve quadratic equations (or at least they would lead today's problem-solver to such), they use concrete numbers instead of general coefficients, and they present the solution in the form of verbal prescription for finding the unknown. The approach is thus purely operational: the focus is on numerical processes and there is no hint of abstract objects other than numbers.

The *rhetoric algebra* (it is how the verbal algebra is referred to by historians) was practiced from the earliest times till the sixteenth century. This is also the kind of algebra encountered by today's school children well before any formal notation is introduced. Naturally, what children are expected to solve 'rhetorically' is much simpler and the words in which they put their solutions sound differently; still, it is basically the same type of mathematics: verbal and operational.

At this point it must be stressed that the operational character of algebra is not inseparable from its being verbal. It is true that as long as algebraic ideas are dressed in words and in words only, it is difficult to imagine the more advanced structural approach, where the computational processes are considered in their totality from a higher point of view, and where operational and structural slants meet in the same representations. To put it differently, words are not manipulable in the way symbols are. It is this manipulability which makes it possible for algebraic concepts to have the object-like quality. It is the possibility of performing higher-level processes on the processes represented by compact expressions that spurs structural thinking. Thus, introduction of a symbolic notation seems *necessary* for reification. On the other hand, it is *not sufficient* for the transition to the structural mode. As already stated in our first example, operational conceptions may also be conveyed through symbolic representations. In any case, the verbal means perpetuate operational thinking. This may be one of the reasons why the absolute reign of the operational algebra lasted thousands of year.

2.2. *Algebra as Generalised Arithmetic: the Structural Phase*

Before we make a further step into the history of algebra, we should make some preliminary theoretical clarifications.

First, it is important to explicitly stress an important implication of the closing remark in the last section: the history of algebra is not a history of symbols. True, from a certain stage on, algebraic concepts become practically inseparable from symbols just like the artist's conception of a picture or of a sculpture is inseparable from its physical embodiment. Indeed, the basic concepts of modern algebra can hardly be conveyed by any means other than algebraic symbols. Moreover, new algebraic knowledge is constructed through manipulations and investigations of formal expressions, and therefore the changes in symbolism parallel conceptual metamorphoses. Thus, from the moment modern algebraic notation was introduced, the history of algebra and the history of symbols, although certainly different from each other, became so intimately intermingled that it is practically impossible to tell the story of one of them without telling the story of the other.

Another important issue which should be explicated at this point regards the correlation between the advancement from operational to structural algebra and the difficulty of demands which the algebraic ideas place on their user. The point we wish to make is that climbing the hierarchy of algebraic ideas is not necessarily tantamount to increase in the sophistication of a person's thinking. One may even feel tempted to say that the opposite is the case: the transition from operational to structural algebra, although a significant step forward in the degree of abstraction and generality, results in facilitating the performance rather than in adding complexity. Although reification itself may be difficult to achieve, once it happens, its benefits become immediately obvious. The decrease in difficulty and the increase in manipulability is immense. What happens in such a transition may be compared to what takes place when a person who is carrying many different objects loose in her hands decides to put all the load in a bag. To fully appreciate the facilitating impact of reification (attained through an appropriate symbolization), it is enough to have a glimpse at the samples of rhetoric algebra presented in Table I. The algorithms which seem so easy and obvious when performed through formal manipulations on concise formulae become very intricate when dressed in verbal, purely operational, representation. Thus, the medieval mathematicians deserve our highest esteem for systematically dealing with problems as advanced and complex as cubic and quartic equations without the ingenious apparatus of structural symbolic algebra (the complex algorithms were presented in a rhetoric manner by Cardan in his famous *Ars Magna* of 1545). As far as their capabilities and the sophistication of thinking are concerned, they can only be compared to the best of today's mathematicians dealing with the most advanced problems of modern mathematics. We should keep all these facts in mind later, when focusing on issues of individual learning and problem solving.

2.2.1. *Algebra of a fixed value (of an unknown).* As previously stated, algebraic symbolism is unrivalled in its power to squeeze the operationally conceived ideas into compact chunks and thus in its potential to make the information easier to comprehend and manipulate. If introduced earlier, the parsimonious notation could have changed the rate of algebra's development, something that the science of computations seems to have needed badly since the day it was born. In comparison to geometry, where the means for structural thinking in the form of graphic representations were readily available, the progress of algebra was slow and hesitant. By the end of the sixteenth century algebra approached such a degree of complexity that without a transition to a structural mode its further development would have been stymied. Historians of mathematics have often wondered why the thinkers of the past, having such strong incentives for a radical change of the method, did not come across the idea of non-verbal representations much earlier.

Although so natural to us, to them the concept of symbolic notation was evidently not at all obvious. In fact, the difficulty lies probably not so much in the idea of using letters instead of numbers and operations (these appeared from time to time even in ancient writings), as in the necessity to imbue the symbolic formulae with the double meaning: that of computational procedures and that of the objects produced. In arithmetic it is easy to keep these two meanings separate by putting them in different expressions: $2 + 3$ denotes the operation, 5 is the outcome. No such separation is possible in algebra, in an expression like $a + b$ or $3(x + 5) + 1$. Here, the process cannot be actually formed; no added value results from the operations. The formula, with its operational aspect salient (it contains 'prompts' for action in the form of operators), must be also interpreted as the product of the process it represents. Even our most abstract thinking, however, is shaped by metaphors provided by sensory experience (Lakoff and Johnson, 1980), and this experience speaks with force against the idea of a process which produces no added value and ends up being treated as its own product. Indeed, nothing like that is possible in real life: we just cannot eat a *recipe* for a cake pretending it is the cake itself (even though we can *imagine* the cake or ourselves eating the cake)! Thus, our intuition rebels against the operational-structural duality of algebraic symbols, at least initially. (The disbelief with which new kinds of numbers were invariably greeted throughout history is another example of a phenomenon which may probably be ascribed to the same ontological dissonance: such objects like $3/4$, -2 or $\sqrt{-1}$ were born out of operations of division, subtraction, and square-root extraction which did not seem to produce anything at all.)

True, once we manage to overcome this difficulty, it is quickly forgotten. To those who are well versed in algebraic manipulation (teachers among them), it may soon become totally imperceptible. Our eyes are easily blinded by habit and by our own ontological beliefs. Nevertheless, much evidence for the difficulty of reification may also be found in today's classroom, provided those who listen to the students are open-minded enough to grasp the ontological gap between themselves and the less experienced learners. In sections 3.2 and 3.3 we shall substantiate this claim with many examples.

Historical facts indicate that the idea of operational-structural duality was also difficult for generations of mathematicians. It was probably Diophantus (c. 250 A.D.) who made the first significant step in the direction of a structural approach to computational procedures. By systematically intermingling letters with words he created for himself a special brand of algebra, known as *syncopated*. While solving word problems, he constructed such expressions as $10-x$ and $10+x$ (in fact, he wrote equivalent strings of Greek letters) and manipulated them as if they were genuine numbers (e.g. he multiplied them obtaining $100 - x^2$; see Fauvel and Grey, 1987, p. 218). The fact that thirteen centuries after Diophantus mathematicians still preferred the awkward verbosity of rhetoric algebra bespeaks the inherent difficulty of his way of thinking.

Diophantus did not go beyond the use of algebraic expressions in which a letter denoted an unknown but fixed value, and where the resulting expressions represented the numbers obtained by combining the unknown with other numbers. We shall say that what he developed was algebra *of a fixed value*, as opposed to *functional algebra*, where letters represent changing rather than constant magnitudes. The idea of a letter as variable – as a symbol instead of which any number may be substituted – so obvious to us nowadays, never occurred to Diophantus. To demonstrate his solution to a problem as, say, "Find two numbers given their sum and product", he used to choose concrete numbers as givens. It seems that the idea of an algebraic expression as a representation of the final result is something quite different – and more difficult to accept – than using formulae as temporary representations of manipulations on the unknown. It certainly requires a full-blown structural view of algebraic expressions – the ability to relate to the above parametric formula as if it was a number and not just an operation which could not be implemented.

2.2.2. Functional algebra (of a variable). It was not until the sixteenth century that algebraic expressions came to denote functions rather than fixed values. The breakthrough occurred in several steps, the first of which was introduction of special symbols for operations and relations, followed by the idea of a letter as a parameter (a given).

The French mathematician François Viéte (1540–1603) was the first to replace numerical givens with symbols. This invention led to a far-reaching conceptual change in algebra. First, the process-product duality of an algebraic expression almost imposed itself on the mathematicians since it was a part and parcel of the idea of using letters as unspecified numbers (operations on letters, say $3(x+5)+1$, could not be actually implemented, so in order to proceed and do something to the resulting number one had no choice but to refer to the formula as if it stood also for the product of the computation). Second, once the letteral formulae were accepted as representing certain objects, a formal algebraic calculus was created which, among other things, specified the ways equations should be manipulated to be solved. This was a drastic change in comparison to the operational algebra, where problems were solved mainly by reversing computational processes. Third, after the new invention was transferred (mainly by Descartes and Fermat) to geometry to serve as an alternative to the standard graphic representations, and then applied in science (by Galileo, Newton, and Leibniz, among others) to represent natural phenomena, algebra was ultimately transformed from a science of constant quantities into a science of changing magnitudes. By that time, the quest for logical foundations of algebra began. The meaning of algebraic expressions and of their symbolic ingredients was found to be elusive and hard to capture into a mathematical definition. Such names as 'generalized number' or 'variable number' were soon disqualified as lacking precision (see, e.g., Frege, 1970). The problem was eventually solved by abandoning the idea of defining the variable as such and by offering instead an interpretation for an algebraic formula in its totality. *Function*, a new kind of abstract mathematical object, was created to serve as a referent for such expressions as $3(x + 5) + 1$ or $x^2 + 2y + 5$.

The problematic nature of the new concept, noted and analyzed in detail by historians and by psychologists (see, e.g., Kleiner, 1989; Dubinsky and Harel, 1992), is a separate theme on which we will not elaborate in the present paper (another article in this volume is devoted exclusively to this issue). As it has much bearing on our subject, however, an understanding of the inherent difficulty of the notion of function is necessary for those who wish to have a deep insight into the process of learning algebra.

2.3. *Abstract Algebra: Algebra of Formal Operations and Algebra of Abstract Structures*

Abstract objects, such as different kinds of numbers or functions, emerge at these junctions in the development of mathematical knowledge where some new processes are introduced, which are to be applied to certain other, already well-known, processes. An abstract object mediates between the two: it may be viewed as a product of the lower-level process and it lends itself to the higher-level manipulations. Thus, with respect to a given object, these two, lower- and higher-level, processes may be called, respectively, *primary* and *secondary*. For example, the idea of rational number originates in dividing integers by integers (primary process), but an entity like 3/4 fully crystallizes as a number in its own right only when the rules of manipulating it and combining it with other numbers (secondary processes) are established. In algebra, primary processes are the arithmetical operations on numbers, the secondary processes are those for which these numerical operations serve as inputs. The latter kind of processes expresses itself in manipulations on algebraic formulae. Thus, the idea of function constitutes a conceptual bond between numerical calculations and formal algebraic manipulations. It acts as a link through which new algebraic knowledge is tied to the system of arithmetical concepts.

After the Vietean type of algebra established itself as the leading tool for doing mathematics, the next step was to climb to a higher point of view from which the present secondary operations – those performed on functions and expressed in manipulations on formulae – could be watched in their totality and become an object of a systematic study (this is a typical development: a process which is regarded as secondary at one level will become primary at a higher level).

This stage in the development of algebra began in Britain in the third decade of the nineteenth century. Although from now on the story goes well beyond school algebra, it is worth telling for reasons which will become clear when today's students' views on symbolic formulae and equations will be analyzed (section 3.3).

Until the nineteenth century, algebra was regarded as a "universal arithmetic" – a discipline which specialized in expressing in a general way the rules governing arithmetical procedures. However, this interpretation greatly limited the scope and the force of the operations on algebraic formulae (e.g., a restriction $a > b$ was a necessary supplement to the expression $\sqrt{a - b}$). Now, when the focus of attention was shifted to the formal manipulations themselves, mathematicians felt an urge to set algebra free from any confinements. A group of British formalists (A. de Morgan, G. Peacock and D. F. Gregory) suggested that algebra should be relieved from the ballast of its original interpretation. From now on, an algebraic formula should be treated as a thing in itself, interpretable in many different ways but devoid of a meaning of its own. The algebraic expression became an empty vehicle waiting to carry an arbitrary semantic load. The formalist school was interested not so much in the potential 'cargo' as in the rules that governed the movements of the vehicles. As Gregory put it, algebra was to become a science "which treats the combinations of operations defined not by their nature, that is by what they are or what they do, but by the laws of combinations to which they are subject" (Gregory, 1840; as quoted by Novy, 1973, p. 194). Here, the word *operations* is used to denote primary processes, while *combinations* are clearly the secondary processes. Thus, the British formalists initiated a new, higher-level operational stage in algebra. It was the first step in the development of *abstract algebra*.

Although the story of algebra does not end here, it is where our historical account stops. The science of abstract structures such as groups, rings, fields or ideals, initially developed in the nineteenth century, does not belong to our subject as it is not taught at secondary level. Just to complete the picture let us remark that with the advent of group theory, a new structural phase began – a natural successor to the operational higher-level algebra of the British formalists.

The numerous stages in the development of algebra are sumarized in Table II. The scheme reinforces the claim that was made at the very outset: algebra is a hierarchical structure in which what is conceived operationally at one level must be perceived structurally at a higher level. Understanding the nuances of the different interpretations of algebraic expressions and their mutual relations is very important for our further discussion, in which the learning of algebra by today's school children will be analyzed.

3. THE DEVELOPMENT OF ALGEBRAIC THINKING – PSYCHOLOGICAL PERSPECTIVE

3.1. *Preliminary Remarks: Competence in Algebra as a Function of Versatility and Adaptability in the Interpretation of Symbols*

As we have shown in the above historical account, the development of the long sequence of possible approaches to algebra and to its symbolic constructs took thousands of years. Today, to solve one little problem from a standard textbook, the learner must often resort to all the different perspectives together. The example in Table III shows the meandering route through diverse outlooks that one has to take to solve the parametric equation discussed in the introduction to this paper. The problem-solver oscillates between the operational and structural approach, and between one structural interpretation to another.

Some 'real life' examples of such process-object swinging in algebraic problem solving may be found in Moschkovich et al. (1992). Gray and Tall (1991) point to a similar phenomenon in arithmetic. Mason (1989) notes the frequent occurrence and the importance of the "delicate shift of attention from seeing an [algebraic] expression *as* an expression of generality, to seeing the expression *as* an object or property". All the researchers agree than the "flexibility [of the perspective] is a hallmark of competence" (Moschkovich et al., 1992). The reason for this, as suggested by the theory of reification, may be summarized as follows (for a

TABLE II

Stages in the development of algebra

Type	Stage	New focus on	Representation	Historical highlights
1. Generalized Arithmetic	1.1. Operational	1.1.1. Numeric **computations**	Verbal *(rhetoric)*	Rhind papyrus, c. 1650 B.C.
			Mixed: verbal+symbolic (*syncopated*)	Diophantus, c. 250 A.D.
	1.2. Structural	1.2.1. (Numeric) **product** of computations ('*algebra of a fixed value*')	Symbolic (letter as an **unknown**)	16th century, mainly Viete (1540–1603)
		1.2.2. (Numeric) **function** ('*functional*) algebra')	Symbolic (letter as a **variable**)	Viete, Leibniz (1646–1716), Newton (1642–1727)
2. Abstract Algebra	2.1. Operational	**Processes on symbols** (combinations of operations)	Symbolic (no meaning to a letter)	British formalist school (de Morgan, Peacock, Gregory), since 1830
	2.2. Structural	**Abstract structures**	Symbolic	19th and 20th century: theories of groups, rings fields, etc., linear algebra

much more comprehensible treatment of this issue, see Sfard, 1987, 1991, 1992). The operational way of thinking dictates the actual actions to be taken to solve the problem at hand, while the structural approach condenses the information and broadens the view. The abstract objects serve as landmarks with the help of which the problem-solving process may be navigated. Since a jump from operational to structural mode of thinking means a transition from detailed and diffuse to general and concise – from the foot of a mountain to its top – it is only natural that it is accompanied by an increase in student's ability to cope with the task at hand.

Now, let us take a closer look at the notion of flexibility purported to be the source of competence. This particular trait of algebraic thinking seems to be a function of two parameters: the *versatility* of the available interpretations,

TABLE III

Oscillating between approaches when solving an algebraic problem

The problem: For what values of the parameters p and q the equation
$(p + 2q)x^2 + x = 5x^2 + (3p - q)x$ holds for *every* value of x?

A possible solution

A step in the solution (decision, operation)	The approach applied
(1) Each of the component formulae represents a family of quadratic functions. The task is to find these members of the two families that are equal to each other. Two polynomial functions are equal if the coefficients of the same powers of x are equal. Thus, to answer the question we have to solve the system of equations: $$p + 2q = 5$$ $$3p - q = 1$$	(1) Here, the formula is interpreted as representing a **family of functions** (one may think about two **parabolas** and ask for what p and q these two curves overlap)
(2) Let's start with solving the first equation with respect to p: $$p = 5 - 2q$$ $$3p - q = 1$$	(2) *First interpretation* $p + 2q = 5$ is seen as **a string of symbols** to be manipulated according to rules *Second interpretation* $p + 2q$ is **a number**; subtraction of $2q$ (also a number) from $p + 2q$ and 5 preserves the equality.
(3) Let's substitute $5 - 2q$ instead of p in the second equation $$p = 5 - 2q$$ $$3(5 - 2q) - q = 1$$	(3) $5 - 2q$ is treated as **a number** (a product of the process which it represents)
(4) Let's solve the second equation with respect to q: $$\begin{aligned} 3(5 - 2q) - q &= 1 \\ 15 - 6q - q &= 1 \\ 15 - 7q &= 1 \\ -7q &= -14 \\ q &= 2 \end{aligned}$$	(4) the formulae are seen as **strings of symbols** to be subjected to formal operations, according to the rules
(5) Let's substitute 2 instead of q in the first equation and compute the value of p $$p = 5 - 2 * 2 = 5 - 4 = 1$$	(5) the expression $5 - 2q$ which turned into $5 - 2 * 2$ is interpreted as **computational process**
(6) Let's formulate the answer: (the functions) $(p + 2q)x^2 + x$ and $5x^2 + (3p - q)x$ are equal iff $p = 1$ and $q = 2$	(6) Back to the **functional** outlook

and the *adaptability* of the perspective. These two parameters appear to be quite independent. In certain circumstances a person may display his or her ability to see an expression as a process, in another context he or she may view it as the product of this process, and in still another situation as a function. One would say, therefore, that the versatility of his or her outlook is quite impressive. This, however, does not necessarily mean that the person will always be able to adapt the perspective to the task at hand. Although such adaptation would sometimes occur as smoothly and imperceptibly as the accommodation of an eye to a changing perspective, in certain circumstances it may be equally difficult as alternating between different perceptions of a cube represented in a two-dimensional picture. For instance, one may be well aware, in principle, that such expressions as $(p + 2q)x^2 + x$ and $5x^2 + (3p - q)x$ can represent functions, but this particular interpretation would not occur to him or her spontaneously at a time of solving the problem presented in Table III. Thus, pointing to the potential versatility of student's conceptions is not enough to arrive at a good assessment of their algebraic competence. Equally important, the adaptability of their outlook should be tested.

It is one of our basic theoretical assumptions that the assortment of perspectives available to the student grows gradually, roughly following the logical-historical path presented in Table II. The hierarchical structure of algebra and of its different interpretations makes this conjecture fairly plausible. A direct jump, say, over the wide gap separating functional algebra from operationally interpreted algebra may end in broken bones. In the discussion that follows, we shall focus our attention on two critical junctions in school algebra: first, we shall consider the transition from purely operational conception of a symbolic formula to the dual process-product interpretation (from cell 1.1.1. to 1.2.1. in Table II); second, we will investigate the passage from here to the functional approach (to cell 1.2.2.). We shall make an effort to find out how difficult these two steps are for the pupil and what phenomena may be regarded as symptoms of such difficulties. In the last section, the looking glass will be turned at the final outcome of schooling: the question of versatility and adaptability of the conceptions with which the students typically leave the school will be addressed.

3.2. *Toward the Structural Outlook*

Although past investigations did not originate in one consistent conceptual framework, their findings, when combined together and re-examined through the lenses of the theory of reification, will often lead to new insights and to a more comprehensive picture of learning. To make these insights even more powerful, we shall reinforce them with samples of interviews and observations carried out by ourselves with children of various ages and competences. The interviews focused on the notion of propositional formula (equation, inequality). Our interviewees included

Group 1:   Six seventh-graders (age 12–13) of an average and slightly over-
            average ability who, by the time we met them for the first time, were
            already acquainted with the notion of an algebraic expression but not
            with the concept of equation;

Group 2:   Four ninth-graders (age 14–15) of above-average ability who were
            supposed to be well versed in basic algebra, including linear and
            quadratic equations and linear inequalities, and were familiar with the
            notion of function in general and with linear functions in particular;

Group 3:   Four tenth-graders (age 15–16) of above-average ability who had had
            a long experience with algebra in many different contexts, including
            analytic geometry and calculus (thus, could be expected to be well
            acquainted with the functional approach).

There was also another kind of inquiry. Each of the children from Group 1, after
he or she was taught to solve linear equations of the type $ax + b = cx + d$, was
asked to help in explaining the subject to his or her peer for whom the whole
issue of equations was still completely new. In this way, we hoped to get another
perspective on the children's thinking. We believed that since the necessity to
convince an uninitiated is a strong motivational force, listening to children giving explanations to other children may be a more powerful method of inquiry
than asking direct questions. All the interviews and the teaching meetings were
recorded either on video- or on audio-tape.

It must be emphasized that the present paper is by no means a systematic report
of the above research, nor an attempt at a comprehensible presentation of all the
results. The large-scale study from which our interviews were but a small part will
be summarized elsewhere. Here we shall only avail ourselves of those selected
samples which bring our message with particular clarity.

3.2.1. *First step: toward algebra of a fixed value (recognition of a process-product
duality).* Since the transition from purely operational outlook to the dual, process-
product interpretation of algebraic formulae occurs in close vicinity to the point at
which arithmetic meets with algebra, much data relevant to our topic may be found
in the research devoted to this crucially important junction. In the following brief
summary, enriched with our own observations, we shall try to find out how the
transition from operational algebra to structural algebra of a fixed value expresses
itself in student's behavior. With this goal in mind, we shall analyze pupil's initial
understanding of formulae, of the equality sign, and of equations.

It is one of our basic theoretical theses that an operational conception naturally
precedes a structural. Findings collected by ourselves and by other researchers
gave this supposition strong support: it seems that even without direct intervention
from a teacher, the learners initially interpreted algebraic expressions as computational processes. The following are only a few out of many examples that may
be brought to illustrate this point.

Those who listen carefully to the language used by the pupils will usually
notice that the way many children initially refer to algebraic expressions indicates

their operational outlook. It may be easily seen in the interviews quoted by Booth (1988, p. 21). For instance, a child asked by an interviewer to write down the length of a space-ship's path composed of y 11-light-years long segments said: "What, shall I write what I would do?"; and after she eventually contrived the formula $11 * y$, she exclaimed to the interviewer: "What, is that all it was? Why didn't you say so? I thought you wanted an answer." Thus, for this child the expression was a mere prescription for the sought-for quantity, not the quantity itself. In our own teaching experiment with children of Group 1, such references to algebraic expressions as 'this is an exercise that must be done' were noted time and again even when a pupil was explaining what this 'thing' (say $8x$) was which he or she decided to subtract from both sides of an equation. The following fragment of a dialogue between the interviewer (I) and the Group-1 pupil Ayala (A) is particularly enlightening, as it indicates that the operational approach to algebraic formulae is inherited from arithmetic. Ayala was trying to explain how her friend Irit solved an equation.

I: How did Irit go from here [$15x = 8x + 35$] to here [$7x = 35$]?

A: She subtracted an exercise, 8 times x, and she subtracted it also from the other side of the equation.

I: What do you mean when you say 'exercise'?

A: 8 times x is an exercise, it is something you must do. She takes off this exercise. It's like when you have $1 + 2 + 4 = 3 + 4$. Then you can take off 3, and then at the other side you take off 'one plus two'.

I: $1 + 2$ is an exercise and 3 is not an exercise?

A: 3 is a number, it's a result of an exercise. This [points to $1 + 2$] is the exercise: and $8x$ is an exercise and $15x$ is an exercise. We subtract the same exercise from both sides so that what is left is the same.

(Ayala's statements indicate that she is somehow half-way between operational and structural outlook: she already manipulates algebraic expressions as if they were objects, but the language she uses is still operational.)

The conviction that a formula is nothing else than a process waiting to be performed may be responsible for what Collis (1974) called the inability to accept the lack of closure – a student's difficulty with complex expressions not followed by an equality sign and not complemented by the 'result' of the computation written on the other side of this sign (see also Chalouh and Herscovics, 1988). This inability may be responsible for the striking result obtained at the British national survey of 15 year olds (as quoted by Bell, 1992): a wide gap in students' performance was noted in the following, seemingly not too different, problems: 'What is x if $2x + 7 = 45$?' and 'If $A = L * B$ tells us how to work out A, what formula tells us how to work out L?' The success rates on these two questions were 73% and 39%, respectively. The dramatic difference may probably be ascribed to the fact that in the second problem, in which some of the letters played the role of parameters ('givens'), the final result (the value of L) would have to be presented by means of a formula and not of a number. This must have seem unacceptable to those for whom an algebraic expression was still only a process (as it probably was even for Diophantus, who did not refrain from using formulae to present intermediary calculations, but would not use them for the final outcome). Similar difficulty was experienced by Gay, a 15 year old pupil (group 2), when he was trying to simplify the expression $kx - x = -2$. Although Gay had already two years of algebra behind him, and he was regarded by all his past and present teachers as talented in mathematics, he could not cope with the problem. No idea seemed to come to his mind.

I: $kx - x$ – can't you present it in a different way?

G: No. There is a multiplication here, kx, so what can I do?

I: And if I wrote $3x - x$, would you be able to do anything?

G: $3x - x$? It's $2x$.

I: So? Isn't $kx - x$ similar?

G: But this ... but this does not work ... I don't know the value of k.

I: So what?

G: So what can I write? $k - x$?

I: What have you done here [$3x - x$] to get $2x$? What did you do to 3?

G: I subtracted 1.

I: So?

G: So what? Shall I take one off? I don't know ... If I subtract 1 from k I'll be left with the same mess. As if it was ... See, I don't know how to write it.

I: Here [$3x - x$] you subtracted one from 3 and multiplied x by the result, right? Here [$kx - x$] you subtract one from k and...

G: Multiply by x ... But how do I subtract one from k? How do I write it? $k - 1$?

Not surprisingly, when algebraic expressions are seen as processes rather than objects, the equality sign is interpreted as a 'do something signal' (Behr et al., 1976; Kieran, 1981) and not as a symbol of a static relation. The expression on the left-hand side is a process, whereas the expression on the right-hand side must be a result. Once again, the idea seems to come from arithmetic where the sign '=' is used as a prompt for the implementation of a 'program' appearing to the left of this sign. Nowadays, this outlook may get additional reinforcement from the fact that this is exactly the way we use the '=' key in hand-held calculators. It serves here as a 'run' command. When treated in this way, the equality symbol looses the basic characteristics of an equivalence predicate: it stops being symmetrical or transitive. Indeed, young children seem to have no qualms about solving word problems with the help of a chain of non-transitive equalities. For instance, when asked 'How many marbles do you have after you win 4 marbles 3 times and

2 marbles 5 times?', the child would often write:

$$3 * 4 = 12 + 5 * 2 = 12 + 10 = 22$$

(see also Vergnaud et al., 1979). The 'one-way', non-symmetric approach to the equality sign, so symptomatic of the operational perspective on algebra manifested itself with force in the following little incident between a Group-1 pupil Danna and her friend and 'student' Zohar. After Zohar had successfully solved the equation $7x + 157 = 248$, she looked baffled and stymied when presented with the next example, $112 = 12x + 47$. Whereas the observer was puzzled, Danna immediately suggested the reason for Zohar's helplessness. "She doesn't know what to do because the order of the equation confuses her, it's not like it should be", she said.

The spontaneity of the operational outlook is exhibited in the easiness with which many young children handle simple linear equations of the form $ax + b = c$. As was noted more than once in different studies (e.g., Kieran, 1988, 1992; Filloy and Rojano, 1989), for young pupils it is often intuitively obvious that in order to solve this kind of problem, one must just 'undo' what had been done to the unknown. We have witnessed many examples of such spontaneous reversal of the computation in our first interviews with Group-1 pupils. The following little story is one of them. When Snir came to the interview, he knew nothing about equations. The interviewer mentioned to him that in an equation like $7x + 157 = 248$, $7x$ means '7 times x' and without any further explanations asked for a solution. Snir immediately concluded: "Here, I have to find a number so that 7 times this number plus 157 is 248. First, 248 minus 157 is 91. Now, the number times seven ... 91 divided by 7 is 13. The number is 13." The operational character of Snir's algebra is underlined by its rhetorical presentation.

All the above observations point in the same direction: the operational outlook in algebra is fundamental and the structural approach does not develop immediately. Moreover, as we have shown in our historical outline and reinforced with theoretical argumentation, there is an inherent difficulty in the idea of process-object duality – of a recipe which must also be regarded as representing its own product. This difficulty cannot be expected to disappear without some struggle.

Davis (1975) was probably one of the first writers who realized the significance of what came to be known in literature as the 'name-process dilemma' (in fact, the term 'process-product dilemma' seems more adequate, as there is no indication that the student distinguishes between the name of a thing and the thing itself; incidentally, the inability to sever a sign from the signified may be one of the reasons why the duality of algebraic expressions is sometimes so difficult to grasp). Davis pointed out to what until now might have well gone unnoticed by the majority of teachers: the idea of duality is not self-evident and may be perplexing for a student.

A particularly convincing symptom of this difficulty may be the phenomenon of 'didactic cut' in learning to solve equations, noted by Filloy and Rojano (1985, 1989) and confirmed by others (e.g., Herscovics and Linchevski, 1991, 1993).

These researchers discovered that, whereas the solution of an equation of the form $ax + b = c$ is intuitively accessible to most pupils, the equation with an unknown appearing on both sides, such as $ax + b = cx$ or $ax + b = cx + d$, evidently poses a problem. Since in the former equation the equality sign still functions like in arithmetic – operations on one side and the result on the other – they called it 'arithmetical'.

On the grounds of our general claims and assumptions, the apparent difficulty at this particular point in learning is not surprising. The cut runs along the demarcation line between operational and structural algebra. As long as only arithmetical equations were concerned, there was no need to hold the dual process-product outlook. The computational operations and their results remained separated by an equality sign and each side of an equation preserved its particular ontological identity – that of a process and that of an object, respectively. This division of roles is no longer in force in the non-arithmetical equation. The expression on the right-hand side, expected to be a *product* of the left-hand expression, is in fact a process. Without the dual outlook, which would turn this last expression into an object, the equation does not make much sense. This may be clearly seen in the following typical dialogue, taken from the interview with Snir, a thirteen year old pupil of more than average ability (Group 1). Snir, for the first time, is faced with a non-arithmetical equation $15x + 12 = 8x + 47$.

S: One must find something ... that when I multiply 15 by a number and I add 12 to it, it will be equal to eight times a number, and I add 47 to it. Let's start with something simple. Here, times 3 and times 1 [writes 3 over x on the left-hand side and 1 over x on the right-hand side]. No, it's not that ... I don't know.

I: What are you looking for when you are solving the equation?

S: We must find two things. Something that when I multiply it or divide it ... doesn't matter ... will be equal to the other thing. One must find the way of making them equal, there are two equations here.

Much can be learned from this little exchange. It was difficult for Snir to make sense of the equation which looked to him like 'two equations', two processes waiting to be performed while the relation between them was not clear at all (at this point, Ayala said that there are 'two exercises' here). The meaning of the equality sign here, so evident to the experienced person, turned out to be far from obvious for the beginner: what aspects of the two processes should be equal? The object operated upon? The object obtained? Or maybe both? Interestingly enough, for Snir it was obviously the equality of results that was required, whereas the numbers for which this equality would hold could be different. Indeed, he looked for two different values of x on the two sides of equation. We were initially perplexed by this interpretation, and even more so when we noticed the same un-expected approach in almost all interviews with Group-1 children and with their peer-students. The phenomenon seemed surprising because at this stage of learn-ing the pupils already knew the convention that different occurrences of the same letter in a given expression signify the same number. It seems that this principle collapsed in the face of an equality which could not be interpreted on the basis of the previous knowledge (as was stated by Herscovics and Linchevski, 1991, the problem of the double interpretation may be easily solved by an explicit statement of the convention regarding different appearances of the same letter in a given equation; even so, we find it significant that this restriction does not occur to the student spontaneously in spite of former instruction).

The difficulty with non-arithmetical equations becomes even more visible when such an equation must be solved. Here, the technique of 'undoing' no longer works. The structural conception of an algebraic formula is a prerequisite for the comprehension of the strategy that must be used – that of adding, subtracting, multiplying and dividing both sides by the same expression. Indeed, the idea can only be accepted by those for whom the sides of an equation and the expressions with which these sides are operated upon are objects while the equality sign is a symbol of equivalence. That this is not always the case may be seen from the following exchange between Danna and Zohar, the two Group-1 students quoted above:

D: [To solve $15x = 8x + 35$] you subtract $8x$ from both sides now.

Z: But I don't know how much is $8x$, so how can I subtract it? ... I don't even know whether the x's on both sides are equal.

The utterances by Zohar and similar statements by other students leave little doubt as to the source of the difficulty they experienced at this point. The children were not able to relate to a formula as a representation of a ready-made object. For them it was still a process, and how can a process be subtracted from another process? (In fact, Ayala's utterances quoted above hint at a possibility that some children may be able to put up with the idea of 'arithmetic operation' on processes; the question is whether this kind of understanding is really consistent and effective).

The last point to be made in this section regards solving word problems using equations. Equation requires suspension of actual calculations for the sake of static description of relationships between quantities. This approach does not comply with the pruely operational view of algebra.

Moreover, the declarative, structural mode of the algebraic presentation reverses the order in which the operations must be performed if the 'answer' is to be found. These may be the reasons why in many studies, even quite advanced students have been found to prefer verbal-operational mode of solving word problems rather than symbolic-structural way of putting things (Clement et al., 1979; Soloway et al., 1982; Sfard, 1987; Harper, 1987).

Let us end this section with a quotation from Davis (1975):

Many major cognitive adjustments – 'accommodations' rather than 'assimilations' – are required if one is to do the necessary mental flip-flop and start seeing the equal sign in new ways, and even seeing $3/x$ as an 'answer' instead of a problem. It is not entirely clear that this whole new point of view can be acquired by gentle accumulation of small increments. It may resemble the geological phenomenon of earthquake more than the phenomenon of erosion or dust deposit. [p. 29]

The conjecture ventured here by the author is perfectly in tune with the theory of reification: the transition from purely operational to a dual process-object outlook is probably not a gradual smooth movement toward a higher level. Like any reification it is likely to be a quantum leap toward a higher vantage-point.

3.2.2. Second step: toward the functional algebra. The passage from the algebra of a fixed-value (of an unknown) to the functional algebra (of a variable) is not as well documented in the literature as the previous transition – that from purely operational to the dual, process-product, approach. True, much has been written about the notion of function – about the way it develops and about the difficulty with which it is acquired and applied by the majority of people. Although all this is of some relevance to our present subject, it does not provide us with the kind of direct information that is needed to understand the specific problems of the functional approach to algebra. Among the issues that should be addressed are such questions as student's ability to think about algebraic formulae in terms of functions and his or her readiness to apply this outlook whenever appropriate (thus, once again, the problem we are facing is that of versatility and adaptability of pupil's algebraic thinking).

There are several reasons why our present issue is somehow more difficult to cope with than the previous one. For one thing, the significance, or even the very existence, of the particular turning point we are now talking about may be less clear than the former. Even if recognized, it is quite elusive. Whereas it is quite obvious where to look for the transition from the operational to process-product outlook (after all, it is a part and parcel of the passage from arithmetical to symbolic algebra!), it is not that easy to pinpoint the moment in learning when the functional approach becomes truly necessary. Also, distinguishing one functional approach from any other in student's solutions to standard school problems is certainly not a trivial matter.

On top of it, the modern school curricula often introduce the functional approach almost from the beginning, so that, at the face of it, there is not much sense in talking about a *transition* to this approach. The Israeli method of teaching algebra may serve as an example. An advanced structural outlook is assumed here almost simultaneously with the introduction of algebraic symbols. Let us draw a sketchy picture of the method.

Algebraic expressions are introduced before they become a part of an equation or inequality. The 12–13 year old child begins her or his algebraic education with modelling different 'real-life' situations and numeric relationships using symbolic formulae. At this stage he or she is not requested yet to compute any values, just to describe the state of affairs assuming that all the numbers are given. Thus, from the very start, the letters are applied as *variables* rather than as unknowns. *Equations* and *inequalities* are introduced slightly later, as two different, but closely related, instances of a single mathematical notion: *propositional formula* (PF, from now on). This universal construct is defined as 'a combination of symbols (names of numbers, letters, operators, predicates, and brackets) that turns into a proposition when names of numbers are substituted instead of the letters'. The idea of PF is introduced as early as seventh grade, and then equations and inequalities are dealt with simultaneously. Every PF has its *truth-set* (TS, for short), namely the set of all the substitutions that turn this PF into a true proposition. Any two PFs with the same truth sets are called *equivalent*. *Solving an equation or an inequality* means finding its TS. As a consequence of this approach, even the solution procedures are described in set-theoretic terms: to solve, say, an equation E, one must find the simplest possible PF which is equivalent to E. The basic steps which may be taken to transform an equation into an equivalent PF are called *elementary (permissible) operations* (in our language, these are the secondary processes of school algebra).

Although the concept of function is officially introduced some time later (eight grade, sometimes only ninth grade), the above method of teaching provides a good example of a structural, functional approach: letter is presented as a variable, an algebraic expression as a function of this variable, and propositional formulae are interpreted as comparisons between functions. Notably, there is much emphasis on graphical representations of the truth sets, which in the case of an equation with two variables, x and y, often leads to the graph of a function $y = f(x)$. This uncompromisingly structural way of dealing with the subject is certainly very attractive due to its mathematical elegance, consistency, and universality. However, since it is somehow at odds with the epistemological and historical order in which the algebraic concepts seem to be related to each other, one cannot be sure that the functional approach is the best slant to begin with.

In a series of interviews carried out in Group 2 (age 14–15, Grade nine) and Group 3 (age 15–16, Grade ten), it was our goal to assess students' familiarity with the functional approach and their competence in using it in different contexts. In preliminary talks with our interviewees, we tried to ascertain that all of them were beyond the 'didactic cut' and that they had assimilated the fixed-value approach. Indeed, we found that they could cope with many kinds of equations and inequalities and could explain the necessary moves in terms of operations on both sides of a propositional formula (e.g., "Here I added $2x$ to both sides... it is permissible because when I do it on both sides, one thing balances the other and the two sides remain equal").

Our first task was to decide in what kind of problems the functional approach becomes indispensable or at least more helpful than any other outlook. We scrutinized several textbooks and identified three representative examples: a quadratic inequality, a system of equations with an infinite truth set (singular equations) and a system of parametric equations. An explanation for why the functional approach is indeed vital in dealing with each of these problems is given below.

Let us precede the explications and the examples of our findings with a methodological remark. While presenting the interviewee with our three questions, we were well aware that he or she may not yet be acquainted with the techniques for answering some of the problems (e.g., quadratic inequalities). For us, it was an advantage rather than a drawback. Since we were interested in diagnosing the student's ways of interpreting algebraic constructs, we had to prevent his or her conceptions from being buried under mechanical algorithmic behavior, typically displayed in solving standard problems. In theory, even our younger interviewees were endowed with all the information necessary to actually implement even those tasks which were not yet exercised at school. The intriguing question was whether their algebraic knowledge was versatile and adaptable enough to allow them at least to understand the problems. In the following short account of our clinical interviews we will not try to give any statistics or generalizations. We shall confine ourselves to three brief case studies which seem to us most enlightening and quite typical. The fact that all our interviewees were considered by their mathematics teachers as being perceptive and successful makes these three episodes particularly telling.

Problem 1: Quadratic inequality

$$\boxed{\text{Solve:} \quad x^2 + x + 1 > 0}$$

Inequalities were introduced to Alon (age 15) more than a year before we met. By the time of the interview he was already well versed in solving linear inequalities. Thus, for the reasons we explained above, it was clear that presenting him with this kind of problem would not provide us with much information. Like for all Israeli pupils, an inequality was supposed to be for Alon just a special case of a propositional formula, not much different from an equation. To us, however, the former kind of PF seemed more difficult, as it clearly demanded a more advanced structural outlook. Whereas equations may be understood and solved on the grounds of a fixed-value approach, namely by treating a letter as a certain unknown number and each side of an equation as a concrete product of operations on this number, an inequality requires that the values of the component formulae are tested and compared for different values of the letter. Thus, in the inequality, the letter plays the role of a variable and the component expressions – of functions of this variable (it goes without saying that the more basic operational approach is out of question: unlike the equality sign, the symbol '>' cannot be interpreted as a 'do something' signal). Perhaps the most efficient way of solving a quadratic inequality such as the one we have chosen is to look at the graph of the appropriate function (in our case: $x^2 + x + 1$) and choose those segments of the x-axis which run below the graph. Although all the pupils we talked to had some previous experience in drawing parabolas, our expectations as to their performance were not very high. A study by Even (1988) with prospective mathematics teachers has already shown that "relating solutions of equations to values of corresponding functions in a graphical representation" is a tough task with which not many learners can cope. Indeed, not even one of our interviewees, whether the younger or the older, resorted to graphing. Not even one of them solved the inequality. The following dialogue between Alon and the interviewer is quite typical.

(1)   I: [Pointing to the inequality] What is it? What do we call such a thing?

(2)   A: Quadratic equation.

(3)   I: Equation?

(4)   A: No, inequality.

(5)   I: What do we look for when we solve it?

(6)   A: We try to find out what the left-hand side is equal to.

(7)   I: What do you mean?

(8)   A: We check how much greater than zero it is and whether it really is greater than zero.

(9)   I: Could you be more precise? What are you looking for? What do you want to get and to write down in the end?

(10)   A: That $x^2 + x + 1$ ... I want to find x^2 and x ... then I can substitute and check whether it is true, this equation ... this inequality.

(11)   I: Say it again. What are you looking for? Chairs, pens, tables?

(12)   A: A number.

(13)   I: A certain number?

(14)   A: Yes, one number.

(15)   I: The special number that ... what?

(16)   A: That I can substitute and find a solution.

(17)   I: What do you mean by 'find the solution'?

(18)   A: Solution, it means that the inequality is true.

From what Alon is saying it seems that he is 'stuck' in the fixed-value view of algebraic formula. In utterances 6 and 9 he refers to $x^2 + x + 1$ as if it had only one, well defined, value. Statements 10, 12, 14, and 16 also indicate that the letter x is for him just a code name for a certain concrete number. With this approach, Alon could not be expected to be able to cope with the problem. To solve the inequality he turned to some old routines, obviously hoping that they will do the job for him even if he could not tell why.

(19)   I: Do you know how to do what you want to do?

(20)   A: Maybe... [writes $x_{1,2} = (-1 \pm \sqrt{1 - 4 \cdot 1 \cdot 1})/2$]. I have here the square root of -3. There is no solution.

(21)   I: So?

(22)   A: So this [points to the inequality] is not true.

(23)   I: What do you mean?

(24)   A: That whatever I substitute, it will be less than zero, or maybe
               zero, but it won't be more than zero.

Alon was unable to assume the functional approach which would allow him to deal
with the inequality in a meaningful way. This handicap led him to a mechanical
behavior grounded in habits which he never tried to revise. The mere occurrence
of a square root of a negative number was for him a signal that the answer should
be 'no solution' (while, in fact, any number fulfills the inequality!). His response
was given automatically and there was no attempt on his part to postpone the
decision in order to check how it related to the particular contents of the question
he was asked.

Problem 2: Singular system of equations

$$\text{Solve:} \quad \begin{cases} 2(x-3) = 1-y \\ 2x + y = 7 \end{cases}$$

Singularities and the things that happen at the fringes of mathematical definitions
are often the most sensitive instruments with which student's understanding of
concepts may be probed and measured. This is also the case with regard to
singular simultaneous equations.

Without a functional approach to algebraic formulae, one is not likely to realize
that a system of linear equations may have infinitely many solutions. If the letters
in the equations represent unknown but fixed numbers, how can anybody expect
that one or both of these fixed numbers will be 'any number'? Moreover, in the
case of two linearly dependent equations, the truth set itself is, in fact, a function:
to each value of x there is a corresponding value of y. To be prepared for this
possibility, the student must realize that each of the component equations may be
considered as representing a function, and the graphs of the two functions can
coincide in any number of points. Only if he knows this, will he be capable of
interpreting in the correct way the tautological equality, such as $0 = 0$, which is
usually obtained as the final output of the routine solving procedure applied to a
system of linearly dependent equations.

Unlike in the case of quadratic equations, the issue of singular equations
should have been known to all our students, even to the youngest among them.
This subject is usually touched upon almost simultaneously with the introduction
of systems of equations (Grade nine), and it is given additional treatment at more
advanced stages of learning. Even so, some of the interviewees responded to
Problem 2 with such answers as: "The truth set is empty" or "x and y may be any
numbers" (no functional relationship between x and y specified).

Let us take a closer look at the particular way in which the problem was tackled
by 15-year old (ninth grader) Mariella. After several transformations, Mariella
arrived at

$$\begin{cases} 2x + y = 7 \\ 2x + y = 7 \end{cases}$$

(1)   M: This is the same equation. Now, shall I solve it? [Starts humming.] Just a moment, $2x$ ... let's take $2x$ off [Writes: $y = 7 - 2x$.] Let's put $7 - 2x$ instead of y ... [Substitutes in the second equation; writes: $2x + (7 - 2x) = 7$; simplifies and arrives at $7 = 7$.]

(2)   I: So?

(3)   M: So x equals zero [writes $x = 0$]. Do you want y as well?

(4)   I: I don't know, you decide. How do you usually complete such a task in the class? What do you write in the end?

(5)   M: The solution.

(6)   I: Well, so write the solution.

(7)   M: If x is 0 then y is 7. Now, I shall put it into the first equation... No, in the second. So, 2 times 0 is 0 ... [arrives at an equality $7 = 7$]. So my solution is true: (0,7).

(8)   I: Is it the only solution?

(9)   M: Yes.

Instead of constantly monitoring the problem from a more advanced point of view in an attempt to look for meaningful shortcuts, Mariella slips into a less stressful, to her taste obviously more secure, algorithmic mode. At the point where the two equations became identical – the fact that she explicitly noted in (1) – the functional outlook combined with the knowledge of linear functions would have given her an immediate answer. Mariella remains unaware of this simple option and uses the routine substitution method instead.

Similar to Alon, Mariella might be thinking in terms of fixed values rather than of variables and functions. Her moves could be dictated by a tacit assumption that x and y represent specific numbers. Alternatively, she might just be manipulating symbols in a routine way without actually giving any thought to the meaning of the letters. Her expectation to find an expression of the form '$x =$ number' in the end of the process was strong enough to force her into faulty reasoning (in (3), she concluded that $x = 0$ from $7 = 7$). The conviction as to the general nature of the solution made her insensitive to the mistake. She did not change her mind even when the interviewer interrogated her on the reasons for her decision:

(10)   I: How from here [$7 = 7$] did you conclude that $x = 0$?

(11)   M: x was cancelled, so I could cancel the 7. And then 0 times x equals 0.

Problem 3: Parametric equations

> Is it true that the following system of linear equations:
> $$\begin{cases} k - y = 2 \\ x + y = k \end{cases}$$
> has a solution for every value of k?

Our own familiarity with variables often makes us insensitive to the subtle difference between equations with numerical coefficients and those with parameters. For people well versed in solving techniques, the question of whether they are operating on numbers or on letters may seem irrelevant. Thus, even for the teachers, the great conceptual difference between the regular kind of equations and parametric equations would not always be clear. (Incidentally, as it often happens with conceptual subtleties, one glance into history could open their eyes.)

In a problem like the present one, the objects that the student is supposed to consider are not just numbers – they are functions. To understand the question, one must realize that each of the equations, $k - y = 2$ and $x + y = k$, represents a whole family of linear functions (or, to put it in different terms, it expresses a family of infinite sets of ordered pairs of numbers), and that for different values of k the system will yield different pairs of such functions. One of the ways to interpret the conjecture presented in Problem 3 is to say that for none of the pairs, the graphs of the two functions are parallel. The conceptual step that must be taken to reach this interpretation may be higher than even the most experienced teachers would guess.

It should be noted that in our particular problem, the functional approach combined with the knowledge of linear functions, if used not only to decipher the question but also to answer it, could lead to an appropriate argument immediately, without any calculations. To show that the truth-set is never empty, it would suffice to realize that the first equation $(k - y = 2)$ represents a horizontal straight line, while the graph of the second $(x + y = k)$ is oblique. Two such straight lines must meet in a point. An alternative algebraic solution would involve applying the usual solution technique and finding whether some of the operations put any restrictions on the value of k (in this case, there are no such limitations and this explains why the truth set is never empty).

While presenting the problem to our interviewees we had every reason to expect that all of them, and especially the older ones, would at least be able to understand the question. We also had certain hopes that the functional way of proving the claim would occur spontaneously. Indeed, except for the fact that the functional approach to propositional formulae was promoted in the school from the very beginning, all our students were well acquainted with linear functions, with their properties and representations. Moreover, the older pupils had a course on analytic geometry behind them and by the time we talked to them they have been studying calculus for almost a year. In spite of all this, not even one of our interlocutors used their knowledge of linear functions to prove that the existence of a solution is independent of the value of k. Some of them (two from each age group) answered the question using the standard algebraic manipulations. For the others, the problem seemed incomprehensible. We find the following conversation with Dina (age 16, tenth grade) representative and significant.

Dina was helpless when faced with the problem. She asked the interviewer what she was supposed to do. The question was obviously not clear to her at all. After a minute or two of looking at the problem she said, "I am groping in the dark." Here is a fragment of our further exchange with her.

(1)   D: [reads the question silently] "... has a solution ..."

(2)   I: What does it mean "has a solution"?

(3)   D: That we can put a number instead of k and it will come out true.

(4)   I: When we say that the system has a solution for every value of k, what is the meaning of the word 'solution'? Is it a number or what?

(5)   D: Yes, it's a number.

(6)   I: One number?

(7)   D: Yes, it's the number that when you put instead of k, then the system is true.

Obviously, Dina had a difficulty with grasping the meaning of the word 'solution' as it appeared here. One possible interpretation of her utterances 3 and 7 is that, for her, the statement 'the equations have a solution' was equivalent to the claim that 'the equations are true' (utterance 3) – as if all the components of the equations had established values. It was probably on the grounds of verbal hints that she decided that in this problem the focus is on k rather than on x and y. In the further exchange, in which the interviewer insisted on getting a clearer explanation, it soon became evident that Dina was far from being firm in her position. Without qualms or explanations, she extended the focus from k to x and y.

(8)   I: This word 'solution' here, to what does it refer? Solution of what?

(9)   D: Of the equations, $k - y = 2$ and $x + y = k$.

(10)   I: What is a solution of these equations?

(11)   D: When we substitute numbers...

(12)   I: Instead of what?

(13)   D: ... instead of x, y, and k, and it comes out true.

(14)   I: So, once more, what are the solutions we are talking about in this question [points to the words 'has a solution']?

(15)   D: I think ... I think that I need three numbers: x, y, and k.

One possible interpretation is that, similar to Alon, Dina was a captive of the fixed-value approach. She could not raise her sight to a higher vantage point which would help her to realize that the objects to be considered here are functions and not numbers. Seeing these advanced abstract entities through the standard algebraic symbols, let alone operating upon them, was clearly beyond her power. This limitation made her unable to interpret the question in a meaningful, consistent way. It left her confused and helpless. Thus, when asked what she was supposed to look for, she had no choice but to 'shoot at random' with pieces of standard statements which had worked in the past. Needless to say, the way Dina tried to actually solve the problem reflected her confusion: she isolated k from the first equation (wrote $k = 2 + y$) and substituted it in the second ($x + y = 2 + y$). At this point she became stymied and could not make a further move.

Before we end this section it should be mentioned that although none of the pupils quoted above were able to demonstrate the functional approach in the pieces of conversations we chose to present here, two of our three interviewees did display some ability to think in functional terms on other occasions. For example, Dina interpreted the quadratic inequality in a correct way: as a problem of a possibly infinite set of values of the variable which would render the quadratic expression a positive value. Alon solved Problem 3 by manipulating the equations, but seemingly with some understanding.

The message from all this may be presented as follows: the functional perspective was not necessarily beyond the assortment of approaches potentially available to our students. Rather, the problem we have witnessed was that of adaptability: the functional approach was not always accessible; sometimes, even if indispensable, it would not be applied spontaneously.

3.3. *When Something Goes Wrong: Pseudostructural Approach*

Despite the functional approach being promoted in Israeli schools, and in spite of the fact that the students we talked to scored quite high in their achievements in mathematics, what we discovered through the interviews was rather alarming. The usually successful pupils displayed very limited ability when thinking about propositional formulae in the advanced structural terms of functions and truth sets. This fact merits special attention, as a danger may be lurking here of what we called elsewhere *semantically debased* or *pseudostructural* conceptions (see Sfard, 1991, 1992; Linchevski and Sfard, 1991).

Let us clarify these terms. As we explained before, abstract objects, such as functions or sets, play the role of links between the old and the new knowledge (see Fig. 1a). In algebra, function is what ties together the arithmetical processes (primary processes) and the formal algebraic manipulations (secondary processes). Thus, reification of the primary processes, or, in the case of algebra, the acquisition of the structural functional outlook, is a warranty of relational understanding. However, the data we collected up to this point provided sufficient evidence that reification is inherently very difficult. It is so difficult, in fact, that at a certain level and in certain contexts, a structural approach may remain practically out of reach for some students. Once the developmental chain has been broken (Fig. 1b), the process of learning is doomed to collapse: without the abstract objects, the

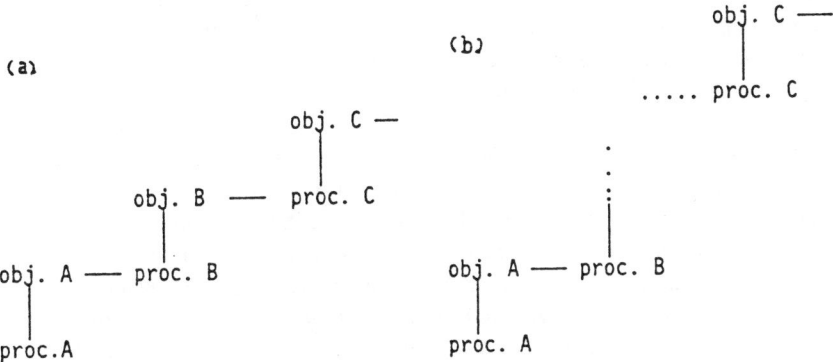

Fig. 1.   Development of mathematical concepts as transitions from operational to structural conception. (a) A 'healthy' developmental chain; (b) a broken developmental chain: process B has not been reified, so there is no link between B and C.

secondary processes will remain 'dangling in the air' – they will have to be executed ... on nothing. Unable to imagine the intangible entities (functions, sets) which he or she is expected to manipulate, the student use pictures and symbols as a substitute: a graph of a function or an algebraic formula, a name of a number, the letters 'ϕ' and x' – each of these signs will turn into a thing in itself, not standing for anything else (how literal such identification may be can be learned from the study by Wagner, 1981, which disclosed that for the majority of secondary school students, a change in the name of a variable leads to a completely new equation). In this case, we shall say that the learner developed a *pseudostructural* conception: he or she mistakes a signifier for the signified. In the absence of abstract objects and their unifying effect, the new knowledge remains detached from its operational underpinnings and from the previously developed system of concepts. In these circumstances, the secondary processes must seem totally arbitrary. The students may still be able to perform these processes, but their understanding will remain instrumental.

In the conversations cited in the last section, there were some disquieting incidents in which one may see symptoms of our interviewees' propensity for a pseudostructural approach. One salient example is the way Alon tried to solve Problem 1 (utterances 19–24). The boy was obviously indifferent to the difference between a quadratic inequality and a quadratic equation, and it was just the form of the left-hand side expression $x^2 + x + 1$ that provided him with clues for his decisions. He applied the formula for the roots mechanically and interpreted the outcome like he used to do when solving equations. Another illustration is Dina's inability to pinpoint the difference between the role of the parameter k and that of the variables x and y, and her confusion as to the meaning of the term 'solution' in the particular context of Problem 3. Both students acted as if they were not aware that the strings of symbols might be interpreted in many different ways, depending on the context. They didn't seem to recognize the existence of any objects external to the letters themselves (except, perhaps, some unknown numbers, but this interpretation soon proved unhelpful).

Both these cases seem indeed to be typical examples of pseudostructural thinking: both students acted as if they were handling some kind of object, but their thinking was completely inflexible and the appropriate kind of structural interpretation was unavailable. That this kind of conceptions are quite widespread was clearly shown in our study (Sfard and Linchevski, 1993), in which 280 secondary-school students (age 15–17) were directly interrogated on the meaning of such algebraic notions as *solving equations, permissible operations, equivalence of equations*. The vast majority of the pupils could not provide any sensible justification for the permissible operations and it was obvious that for them these were no more than arbitrary 'rules of the game'. To be more precise, in the eyes of many respondents, solving equations and inequalities was tantamount to performing a certain algorithm. In this game, an expression of the form '$x = $ number' or '$x > $ number' was but a 'halting signal'.

We have already observed that many pupils seem unable to cope with singular propositional formulae – equations and inequalities in which the variable disappears at a certain stage of the solution process. We illustrated this claim with the case of Mariella (Problem 2). Now a possible reason for the difficulty becomes even more clear. If the student, unable to see the abstract objects behind the symbols, is 'programmed' to regard a problem as solved only when an expression of the form '$x = $ number' or '$x > $ number' is obtained, then in a situation in which such an expression does not appear at all he or she must feel lost and helpless. The following is an excerpt from our interview with 15 year old Naomi which renders these speculations even more plausible.

While solving Problem 2, Naomi arrived at a stage where the first of the two equations turned into $1 = 1$.

(1)	N:	One equals one. It's true, but it gives us nothing. Maybe I shouldn't have opened the brackets [in the first equation] ... I don't know...
(2)	I:	What can you say now about the solution of the system of equations?
(3)	N:	That maybe there is one solution.
(4)	I:	What is it?
(5)	N:	One. The number 1. Or in fact ... in fact I don't think so.
(6)	I:	So?
(7)	N:	There is no solution. An empty set.
(8)	I:	Where do you infer it from?

(9)   N: Because we did all this ... we isolated $2x$, etc. ... to arrive at
the value of y. We substituted this [$7 - y$ instead of $2x$ in the
first equation] and we were left without x, only with y. And
then there was no y as well. Our goal was to find the value
of y and we didn't succeed. So I think that we have here an
empty [truth] set.

In the comment about her lack of success (line 9), Naomi was obviously referring to the fact that she did not arrive at an expected expression of the form
'$y =$ number'. Since her perspective was restricted to the situation in which such
an expression does appear in the end, this 'failure' in the *process* of finding the
solution was interpreted as an absence of the *result* of solving.

4. AFTERWORD: WHERE WE ARE AND WHERE WE SHOULD GO

The episodes discussed in this article give rise to the suspicion that most of the
time algebraic formulae are for some pupils not more than mere strings of symbols
to which certain well-defined procedures are routinely applied. In these students'
eyes, the formal manipulations are the only source from which the symbolic
constructs may draw their meaning.

At face value, this outlook is very close to the view promoted by Peacock and
his colleagues. In fact, the students' conception and the formalist position cannot
be equated, and the differences are certainly more significant than the similarities.
The belief about the nature of the symbolic manipulations is where the formalists
and today's students part. While dealing with symbols, the formalists focus
on *combinations of operations* (Gregory, 1840). The operations that the high-
school student is supposed to master, namely those that can be interpreted as
a generalization of arithmetic calculations, are for the formalists but a point of
departure, mere inputs to the processes they are really interested to investigate.
In other words, the formalist's algebra begins where the school algebra ends.
Besides, although both the mathematician and the pupil view the formal operations
as arbitrary, for the formalist such an approach is a matter of a deliberate choice,
while for the student it is an inevitable outcome of his or her basic inability to link
algebraic rules on the laws of arithmetic.

We started this article with a list of possible perspectives one may assume
while dealing with algebraic constructs. We stressed all along in our discussion
the importance of the versatility and adaptability of student's thinking. The
conclusions we draw are not very encouraging. In a quite consistent way, all
our findings have shown that more often than not, the pupil cannot cope with
problems which do not yield to the standard algorithms. Since it became clear
that the functional approach is not easily accessible even for the better students, it
is not so difficult to understand why the mechanistic, pseudostructural approach
may eventually dominate student's thinking like an overgrown weed, leaving no
room for other, more meaningful perspectives.

As we explained more than once, the sense of meaningfulness comes with the ability of 'seeing' abstract ideas hidden behind the symbols. However, "mathematical objects and structures that the teacher can 'see' are unlikely to be apparent to the student" (Cobb, 1988). On the other hand, it would be a mistake to claim that student's activities and decisions are entirely devoid of an inner logic. With Davis (1988) we shall say that "students usually *do* deal with meanings, and when instructional programs fail to develop appropriate meanings, *students create their own meanings* – meanings that are sometimes not appropriate at all" (p. 9). Although somehow flat and one-sided, the student's do-it-yourself algebra is not devoid of certain consistency. The problem is that those who adopt the pseudostructural approach and confuse the powerful abstract objects with their representations do not realize that the symbols themselves cannot perform the magic their referents are able to do: they cannot glue together lots of detailed pieces of knowledge into one powerful whole. Thus, when talking about the mechanistic, pseudostructural outlook, one may certainly say "though this be method, yet there is madness in it".

The confrontation between the developmental model we promoted in this paper and the structural way of teaching algebra hints at a possible reason for the unsatisfactory results of schooling. The curriculum literally reverses the order in which algebraic notions seem to be related to each other, the order in which they developed through ages. The advanced structural approach is assumed at the outset even though the student is evidently not ready yet to grasp the idea of process-object duality, let alone to cope with the functional outlook. From the very beginning, the letter is supposed to play the role of a variable and not just of an unknown, although we have seen that the former is more advanced than the latter. The stand-alone algebraic formula is introduced before it is incorporated into an equation or inequality, even though it is only in the context of simple ('arithmetical') equations that it may be interpreted in the operational way, more accessible for the young student. Finally, equations and inequalities are taught simultaneously although the latter put heavier demands on student's understanding. The conceptual pyramid is put on its head, so it is only natural that it has a tendency for falling. Revision of the order in which the pupils are exposed to the central ideas of algebra is likely to bring some improvement. It seems reasonable that we can capitalize on the students' natural propensity for an operational approach by beginning with processes rather than with ready-made algebraic objects. As to the latter, there is some evidence that their construction in a student's mind may be helped by the computer (Dreyfus and Halevi, 1988; Waits and Demana, 1988; Schwartz et al., 1990; Breidenbach et al., 1992). (This does not suggest that the early introduction of the functional approach is the only culprit, or that teaching functional algebra from the very start cannot be successful in any circumstances. For example, it is fairly possible that the massive use of computer graphics in teaching functions will reverse the 'natural' order of learning so that the structural approach to algebra will become accessible even to young children. Also, one should remember that no alterations in the organization or presentation of subject matter can suffice to significantly enhance learning. A real change for the better will not come until the teachers find ways to boost students' willingness to struggle for meaning.)

We hope that there is much potential in the suggested didactic ideas and we are now engaged in a teaching experiment in which they are systematically put to test. Notwithstanding our belief that the method may work, there is a question that keeps bothering us: even if our students succeed in acquiring a versatile and adaptable assortment of perspectives, how durable and robust will their flexible knowledge be in the long run? The chances are that pseudostructural thinking may sometimes be compared to secondary illiteracy. After all, the mechanical mode has so much to offer to intellectual sluggards: it exempts the problem-solver from the need for constant alertness, from the strain which inevitably accompanies going back and forth from one perspective to another. As testified by Souriau, mathematicians themselves yield to the charm of mechanized symbolic manipulations:

> Does [the algebraist] follow [the original meaning of symbols] through every stage of the operations he performs? Undoubtedly not: he immediately loses sight of them. His only concern is to put in order and combine, according to known rules, the signs which he has before him; and he accepts with confidence the results thus obtained. [quoted by Hadamard, 1949, p. 64]

The problem is that unlike the mathematician, the student may easily become addicted to the automatic symbolic manipulations. If not challenged, the pupil may soon reach the point of no return, beyond which what is acceptable only as a temporary way of looking at things will freeze into a permanent perspective. When it happens, there is not much chance that the student will be able to explain his or her decisions. If asked for justification, he or she may become as confused as a centipede who has been required to tell how it moves its legs. Thus, to fight pseudostructural conceptions it may be not enough to reform the teaching method in the ways proposed above. It seems very important that we try to motivate our students to actively struggle for meaning at every stage of the learning. We must make them active sense-seekers who, as Davis (1988, p. 10) put it, would 'habitually' "*interpret* situations, *interpret their actions*, think of the *meanings* of symbols and *meanings* of symbols and operations".

ACKNOWLEDGMENTS

We wish to thank Paul Cobb, Tom Kieren, and Uri Leron for their careful reading of the first draft and for helpful remarks.

REFERENCES

Behr, M., Erlwanger, S. and Nichols, E.: 1976, *How Children View Equality Sentences*, (PMDC Technical Report No. 3), Florida State University (ERIC Document Reproduction Service No. ED 144802).

Bell, A.: 1992, 'School algebra – what it is and what it might be', paper presented at the meeting of Algebra Working Group at ICME 7, Quebec, Canada.

Beth, E. W. and Piaget, J.: 1966, *Mathematical Epistemology and Psychology*, D. Reidel Publishing Company, Dordrecht, The Netherlands.

Booth, L.: 1988, 'Children's difficulties in beginning algebra', in A. F. Coxford (ed.), *The Ideas of Algebra, K-12* (1988 Yearbook), NCTM, Reston, VA, pp. 20–32.

Boyer, C. B.: 1985, *A History of Mathematics*, Princeton University Press, Princeton, NJ (originally published in 1968).

Breidenbach, D., Dubinsky, E., Hawks, J. and Nichols, D.: 1992, 'Development of the process conception of function', *Educational Studies in Mathematics* 23, 247–285.

Chalouh, L. and Herscovics, N.: 1988, 'Teaching algebraic expressions in a meaningful way', in A. F. Coxford (ed.), *The Ideas of Algebra* (1988 Yearbook), NCTM, Reston, VA, pp. 33–42.

Clement, J., Lochhead, J. and Soloway, E.: 1979, *Translation Between Symbol Systems: Isolating a Common Difficulty in Solving Algebra Word Problems*. COINS technical report No. 79–19, Department of Computer and Information Sciences, University of Massachusetts, Amherst.

Cobb, P.: 1988, 'The tension between theories of learning and instruction in mathematics education', *Educational Psychologist* 23, 87–104.

Collis, K. F.: 1974, 'Cognitive development and mathematics learning', paper presented at Psychology of Mathematics Education Workshop, Center for Science Education, Chelsea College, London.

Crowe, M.: 1988, 'Ten misconceptions about mathematics and its history', in W. Asprey and P. Kitcher (eds.), *History and Philosophy of Modern Mathematics*, Minnesota Studies in the Philosophy of Science, Vol. XI, University of Minnesota Press, Minneapolis.

Davis, P.: 1975, 'Cognitive processes involved in solving simple arithmetic equations', *Journal of Children's Mathematical Behavior* 1(3), 7–35.

Davis, R.: 1989, 'Research studies in how humans think about algebra', in S. Wagner and C. Kieran (eds.), *Research Issues in the Learning and Teaching of Algebra*, Lawrence Erlbaum, NCTM, Hillsdale, NJ, pp. 266–274.

Davis, R.: 1988, 'The interplay of algebra, geometry, and logic', *Journal of Mathematical Behavior* 7, 9–28.

Douady, R.: 1985, 'The interplay between different settings: Tool-object dialectic in the extension of mathematical ability – Examples from elementary school teaching', in L. Streefland (ed.), *Proceedings of the Ninth International Conference for the Psychology of Mathematics Education*, Vol. 2, State University of Utrecht, Subfaculty of Mathematics, OW&OC, Utrecht, The Netherlands, pp. 33–52.

Dreyfus, T. and Halevi, T.: 1988, 'Quadfun – A case study of pupil-computer interaction', paper presented to the theme group on Microcomputers and the Teaching of Mathematics at ICME 6, Budapest, Hungary.

Dubinsky, E.: 1991, 'Reflective abstraction in advanced mathematical thinking', in D. Tall (ed.), *Advanced Mathematical Thinking*, Kluwer Academic Publishers, Dordrecht, pp. 95–123.

Dubinsky, E. and Harel, G. (eds.): 1992, *The Concept of Function: Aspects of Epistemology and Pedagogy*, MAA Notes, Vol. 25, Mathematical Association of America.

Even, R.: 1988, 'Pre-service teachers' conceptions of the relationship between functions and equations', in Borbas (ed.), *Proceedings of the Twelfth International Congress of the PME*, Vol. I, Vesprem, Hungary, pp. 304–311.

Fauvel, J. and Gray, J.: 1987, *The History of Mathematics – A Reader*, Macmillan Education, London, The Open University, Milton Keynes.

Filloy, E. and Rojano, T.: 1985, 'Operating the unknown and models of teaching', in S. K. Damarin and M. Shelton (eds.), *Proceedings of the Seventh Annual Meeting of PME-NA*, Ohio State University, Columbus, pp. 75–79.

Filloy, E. and Rojano, T.: 1989, 'Solving equations: the transition from arithmetic to algebra', *For the Learning of Mathematics* 9(2), 19–25.

Frege, G.: 1970, 'What is function?', in P. Geach and M. Black (eds.), *Translations from the Philosophical Writings of Gottlob Frege*, Blackwell, Oxford (German original published in 1904).

Freudenthal, H.: 1978, *Weeding and Sowing*, D. Reidel Publishing Company, Dordrecht, The Netherlands.

Garcia, R. and Piaget, J.: 1989, *Psychogenesis and the History of Science*, Columbia University Press, New York.

Greeno, G. J.: 1983, 'Conceptual entities', in D. Genter and A. L. Stevens (eds.), *Mental Models*, pp. 227–252.

Gregory, D. F.: 1840, *On the Nature of Symbolic Algebra*, Trans. Roy. Soc., Vol. 14, Edinburgh, pp. 208–216.

Gray, E. and Tall, D. O.: 1991, 'Duality, ambiguity and flexibility in successful mathematical thinking', in F. Furinghetti (ed.), *Proceedings of the Fifteenth Conference for the Psychology of Mathematics Education*, Vol. 2, Assisi, Italy, pp. 72–79.

Hadamard, J. S.: 1949, *The Psychology of Invention in the Mathematics Field*, Princeton University Press, NJ.

Harel, G. and Kaput, J.: 1991, 'The role of conceptual entities in building advanced mathematical concepts and their symbols', in D. Tall (ed.), *Advanced Mathematical Thinking*, Kluwer Academic Publishers, Dordrecht, The Netherlands, pp. 82–94.

Harper, E.: 1987, 'Ghost of Diophantus', *Educational Studies in Mathematics* **18**, 75–90.

Herscovics, N. and Linchevski, L.: 1991, 'Pre-algebraic thinking: range of equations and informal solutions used by seventh graders prior to any instruction', in F. Furinghetti (ed.), *Proceedings of the Fifteenth Conference for the Psychology of Mathematics Education*, Vol. 2, Assisi, Italy, pp. 173–180.

Herscovics, N. and Linchevski, L.: 1993, 'The cognitive gap between arithmetics and algebra', forthcoming in *Educational Studies in Mathematics*.

Kaput, J. J.: 1989, 'Linking representations in the symbol system of algebra', in S. Wagner and C. Kieran (eds.), *Research Issues in the Learning and Teaching of Algebra*, Lawrence Erlbaum, Hillsdale, NJ, pp. 167–194.

Kieran, C.: 1981, 'Concepts associated with equality symbol', *Educational Studies in Mathematics* **12**(3), 317–326.

Kieran, C.: 1988, 'Two different approaches among algebra learners', in A. F. Coxford (ed.), *The Ideas of Algebra* (1988 Yearbook), NCTM, Reston, VA, pp. 91–96.

Kieran, C.: 1991, 'A procedural-structural perspective on algebra research', in F. Furinghetti (ed.), *Proceedings of the Fifteenth International Conference for the Psychology of Mathematics Education*, Vol. 2, Assisi, Italy, pp. 245–254.

Kieran, C.: 1992, 'The learning and teaching of school algebra', in D. A. Grouws (ed.), *The Handbook of Research on Mathematics Teaching and Learning*, Macmillan, New York, pp. 390–419.

Kleiner, I.: 1989, 'Evolution of the function-concept: A brief survey', *College Mathematics Journal* **20**(4), 882–300.

Lakoff, G. and Johnson, M.: 1980, *The Metaphors We Live By*, University of Chicago Press, Chicago.

Linchevski, L. and Sfard, A.: 1991, 'Rules without reasons as processes without objects – The case of equations and inequalities', in F. Furinghetti (ed.), *Proceedings of the Fifteenth Conference for the Psychology of Mathematics Education*, Vol. 2, Assisi, Italy, pp. 317–324.

Mason, J. H.: 1989, 'Mathematical subtraction as the results of a delicate shift of attention', *For the Learning of Mathematics* **9**(2), 2–8.

Moschkovich, J., Schoenfeld, A. and Arcavi, A.: 1992, 'What does it mean to understand a domain: A case study that examines equations and graphs of linear functions', paper presented at the 1992 Annual Meeting of the American Educational Research Association, San Francisco, CA.

Novy, L.: 1973, *Origins of Modern Algebra*, Noordhoff International Publishing, Leiden, The Netherlands.

Schoenfeld, A. H., Smith, J. P. and Arcavi, A.: 1993, 'Learning: The microgenetic analysis of one student's evolving understanding of a complex subject matter domain', forthcoming in R. Galser (ed.), *Advances in Instructional Psychology*, Vol. 4, Lawrence Erlbaum, Hillsdale, NJ.

Schwartz, B., Dreyfus, T. and Bruckheimer, M.: 1990, 'A model of function concept in a three-fold representation', *Computers and Education* **14**, 249–262.

Sfard, A.: 1987, 'Two conceptions of mathematical notions: operational and structural', in J. C. Bergeron, N. Hershcovics and C. Kieran (eds.), *Proceedings of the Eleventh International Conference for Psychology of Mathematics Education*, Vol. III, Université de Montréal, Montréal, Canada, pp. 162–169.

Sfard, A.: 1989, 'Transition from operational to structural conception: The notion of function revisited', in G. Vergnaud, J. Rogalski and M. Artigue (eds.), *Proceedings of the Thirteenth International Conference for the Psychology of Mathematics Education*, Vol. 3, Laboratoire PSYDEE, Paris, pp. 151–158.

Sfard, A.: 1991, 'On the dual nature of mathematical conceptions: Reflections on processes and objects as different sides of the same coin', *Educational Studies in Mathematics* **22**, 1–36.

Sfard, A.: 1992, 'Operational origins of mathematical notions and the quandary of reification – the case of function', in E. Dubinsky and G. Harel (eds.), *The Concept of Function: Aspects of Epistemology and Pedagogy*, MAA Notes, Vol. 25, Mathematical Association of America, pp. 59–84.

Sfard, A. and Linchevski, L.: 1993, 'Processes without objects – the case of equations and inequalities', forthcoming in a special issue of *Rendiconti del Seminario Matematico dell'Universita e del Politecnico di Torino*.

Soloway, E., Lochhead, J. and Clement, J.: 1982, 'Does computer programming enhance problem solving ability? Some positive evidence on algebra word problems', in R. J. Seidel and R. E. Anderson (eds.), *Computer Literacy*, Academic Press, New York.

Struik, D. J.: 1986, *A Source Book in Mathematics, 1200–1800*, Princeton University Press, Princeton, NJ (originally published in 1969).

Vergnaud, G., Benhdj, J. and Dussouet, A.: 1979, *La coordination de l'enseignement des mathématiques entre les cours moyen 2e année et la classe de sixième*, Institut National de Recherche Pédagogique, Paris.

Wagner, S.: 1981, 'Conservation of equation and function under transformations of variable', *Journal for Research in Mathematics Education* **12**, 107–118.

Wagner, S. and Kieran, C.: 1989, 'An agenda for research on the learning and teaching of algebra', in S. Wagner and C. Kieran (eds.), *Research Issues in the Learning and Teaching of Algebra*, Lawrence Erlbaum, Hillsdale, NJ, pp. 220–237.

Waits, B. K. and Demana, F.: 1988, 'New models for teaching and learning mathematics through technology', paper presented to the theme group on Microcomputers and the Teaching of Mathematics at ICME 6, Budapest, Hungary.

Wheeler, D.: 1989, 'Context for the research on teaching and learning algebra', in S. Wagner and C. Kieran (eds.), *Research Issues in the Learning and Teaching of Algebra*, Lawrence Erlbaum, Hillsdale, NJ, pp. 278–287.

Anna Sfard
The Science Teaching Centre,
The Hebrew University,
Givat-Ram, Jerusalem 91904,
Israel

Liora Linchevski
School of Education,
The Hebrew University,
Mount Scopus, Jerusalem 91905,
Israel

PATRICK W. THOMPSON

IMAGES OF RATE AND OPERATIONAL UNDERSTANDING OF THE FUNDAMENTAL THEOREM OF CALCULUS[†]

ABSTRACT. Conceptual analyses of Newton's use of the Fundamental Theorem of Calculus and of one 7th-grader's understanding of distance traveled while accelerating suggest that concepts of rate of change and infinitesimal change are central to understanding the Fundamental Theorem. Analyses of a teaching experiment with 19 senior and graduate mathematics students suggest that students' difficulties with the Theorem stem from impoverished concepts of rate of change and from poorly-developed and poorly coordinated images of functional covariation and multiplicatively-constructed quantities.

John Dewey once said that theory is the most practical of all things (Dewey, 1929). Theory is the stuff by which we act with anticipation of our actions outcomes and it is the stuff by which we formulate problems and plan solutions to them. It is in this sense that I consider this theoretical investigation of students' calculus concepts to be a highly practical endeavor. Mine is a theoretical paper driven by practical problems. The theoretical side has to do with imagery and operations in the constitution of students' understanding of the Fundamental Theorem of Calculus; the practical side is motivated by our general lack of insight into the poor quality of calculus learning and teaching in the United States.

A primary theme I will develop is that students' difficulties with the Fundamental Theorem of Calculus can be traced to impoverished images of rate. To develop this theme I will need to make several digressions – one to explicate my use of "image", one to explain what I mean by images of rate, and a third to clarify issues surrounding the Fundamental Theorem itself.

IMAGERY AND OPERATIONS

By "image" I mean much more than a mental picture. Rather, I mean "image" as the kind of knowledge that enables one to walk into a room full of old friends and expect to know how events will unfold. An image is constituted by coordinated fragments of experience from kinesthesia, proprioception, smell, touch, taste, vision, or hearing. It seems essential also to include the possibility that images entail fragments of past affective experiences, such as fearing, enjoying, or puzzling, and fragments of past cognitive experiences, such as judging, deciding, inferring, or imagining.[1] Images are less well delineated than are schemes of actions or operations (Cobb and von Glasersfeld, 1983). They are more akin to

Educational Studies in Mathematics **26**: 229–274, 1994.

figural knowledge (Johnson, 1987; Thompson, 1985) and metaphor (Goldenberg, 1988). A person's images can be drawn from many sources, and hence they tend to be highly idiosyncratic.

The roots of this overly-broad characterization of image go back to Piaget's ideas of praxis (goal-directed action), operation, and scheme. I discuss these connections more fully in other papers (Thompson, 1985, 1991, in press a). For the present purpose I will focus on Piaget's idea of an image and its relationship to mental operations.

Piaget distinguished among three general types of images. The distinctions he drew were based on how dependent upon the image were the actions of reasoning associated with it. The earliest images formed by children are an "internalized act of imitation ... the motor response required to bring action to bear on an object ... a *schema* of action" (Piaget, 1967, p. 294). By this I take Piaget to have meant images associated with the creation of objects, whereby we internalize objects by acting upon them. We internalize them by internalizing our actions. Piaget's characterization was originally formulated to account for object permanence. It can also provide insight into a person's creation of mathematical objects (Dubinsky, 1991; Sfard, 1991; Thompson, 1985), and when the development of a person's imagery is halted at this early level it can lead to mathematical understandings that are nothing more than internalized patterns of action (Boyd, 1992).

A later kind of image people create is one having to do with primitive forms of thought experiments. "In place of merely representing the object itself, independently of its transformations, this image expresses a phase or an outcome of the action performed on the object. ... [but] the image cannot keep pace with the actions because, unlike operations, such actions are not coordinated one with the other" (Piaget, 1967, p. 295). It is advantageous to interpret Piaget's description broadly. If by actions we include ascription of meaning or significance, then we can speak of images as contributing to the building of understanding and comprehension, and we can speak of understandings-in-the-making as contributing to ever more stable images.

A third kind of image people come to form is one that supports thought experiments, and supports reasoning by way of quantitative relationships. An image conjured at a moment is shaped by the mental operations one performs, and operations applied within the image are tested for consistency with the scheme of which the operation is part. At the same time that the image is shaped by the operations, the operations are constrained by the image, for the image contains vestiges of having operated, and hence results of operating must be consistent with the transformations of the image if one is to avoid becoming confused.[2]

> [This is an image] that is dynamic and mobile in character ... entirely concerned with the transformations of the object. ... [The image] is no longer a necessary aid to thought, for the actions which it represents are henceforth independent of their physical realization and consist only of transformations grouped in free, transitive and reversible combination... In short, the image is now no more than a symbol of an operation, an imitative symbol like its precursors, but one which is constantly outpaced by the dynamics of the transformations. Its sole function is now to express certain momentary states occurring in the course of such transformations by way of references or symbolic allusions." (Piaget, 1967, p. 296.)

Piaget's ideas of image are similar to those of Kosslyn (1980), and Johnson (1987), but in different degrees. Kosslyn dismisses the idea of images as mental pictures (Kosslyn, 1980, p. 19), characterizing images as highly processed perceptual data that only resembles what is produced during actual perception. On the other hand, Kosslyn's is a correspondence theory, whereby images *represent* features of an objective reality. Piaget's theory assumes no correspondence; it takes objects as things constructed, not as things to be represented (von Glasersfeld, 1978). Also, Kosslyn's notion of image seems to be much more oriented to visualization than is Piaget's. Piaget was much more concerned with ensembles of action by which people assimilate objects than with visualizing an object in its absence. Finally, Kosslyn focuses on images as the PRODUCTS of acting. Piaget focuses on images as the products of ACTING. So, to Kosslyn, images are data produced by perceptual processing. To Piaget, images are residues of coordinated actions, performed within a context with an intention, and only early images are concerned with physical objects.

Piaget's idea of image is remarkably consistent with Johnson's (1987) detailed argument that rationality arises from and is conditioned by the patterns of our bodily experience. Johnson takes to task realist philosophy and cognitive science (which together he calls "Objectivism") in his criticism of their attempts to capture meaning and understanding within a referential framework.[3]

Piaget maintained throughout his career that all knowledge originates in action, both bodily and imaginative (Piaget, 1950, 1968, 1971, 1976, 1980). While Johnson's primary purpose was to give substance to this idea in the realms of everyday life, Piaget was primarily concerned with the origins of scientific and mathematical reasoning – reasoning that is oriented to our understandings of quantity and structure. For example, while Johnson focused on the idea of balance as an image schema emerging from senses of stability and their projection to images of symmetric forces (Johnson, 1987, pp. 72–98), it requires a nontrivial reconstruction to create an image of balance as involving countervailing twisting actions – where we imagine the twisting actions themselves in such a way that it occurs to us that we might somehow measure them. It seems to involve more than a metaphorical projection of balance as countervailing pushes to have an image of balance that entails the understanding that any of a class of weight-distance pairs on one side of a fulcrum can be balanced by any of a well-determined class of weight-distance pairs on the other side of a fulcrum.

I should note that the meaning of "image" developed here is only tangentially related to the idea of concept image as developed by Vinner (Tall and Vinner, 1981, Vinner, 1987, 1989, 1991, 1992, Vinner and Dreyfus, 1989). Vinner's idea of concept image focuses on the coalescence of mental pictures into categories corresponding to conventional mathematical vocabulary, while the notion of image I've attempted to develop focuses on the dynamics of mental operations. The two notions of image are not inconsistent, they merely have a different focus.

The construct of image portrayed here – as dynamic, originating in bodily actions and movements of attention, and as the source and carrier of mental operations – will be fundamental to analyses of students understanding of integral

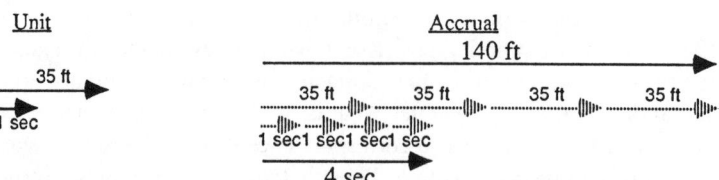

Unit

35 ft

1 sec

Distance and time accrue simultaneously
and in proportional correspondence. One
speed-distance or part thereof is made while
moving for 1 time-unit or corresponding part
thereof. Moving for one time-unit or part
thereof implies moving one speed-distance
or corresponding part thereof.

Accrual
140 ft

35 ft   35 ft   35 ft   35 ft

1 sec1 sec1 sec1 sec

4 sec

Distance and time accrued simultaneously and continuously. Each
speed-distance is a fractional part of the total accrued distance.
Each time-unit is a fractional part of the total accrued time.

Fig. 1. Speed as a rate. Distance and time accrue simultaneously and continuously, and accruals of
quantities stand in the same proportional relationship with their respective total accumulations. This
image supports proportional correspondence, that $\frac{a}{b}$ ths of one accumulation corresponds to $\frac{a}{b}$ ths of
the other accumulation.

and derivative. It will provide the orientation needed to speculate about what the
"something" is that students have in mind when they speak of something changing
or of something accumulating.

IMAGES OF RATE

The development of images of rate starts with children's image of change in
some quantity (e.g., displacement of position, increase in volume), progresses to
a loosely coordinated image of two quantities (e.g., displacement of position and
duration of displacement), which progresses to an image of the covariation of two
quantities so that their measures remain in constant ratio (Thompson, in press a;
Thompson and Thompson, 1992).

The development of mature images of rate involves a schematic coordination
of relationships among accumulations of two quantities and accruals by which the
accumulations are constructed. For example, in the case of constant speed, the
total distance traveled in relation to the duration of the trip can be imagined as
each having accumulated through accruals of distance and accruals of time so that
at any moment during the trip the total distance traveled at that moment in relation
to the total time of the trip is the same as the accrual of distance in relation to the
accrual of time (Figure 1).

Rates which involve time seem to be the most intuitive, but time as a quantity
which can be imagined to vary proportionally with another quantity is a nontrivial construction for students (Thompson and Thompson, in press; Thompson,
in press a). A further abstraction is required to develop an image of rate that
entails the covariation of two non-temporal quantities (e.g., volume and surface
area) and the notion of average rate of change of some quantity over some range
of an independent quantity (e.g., average rate of change of luminance with respect
to the displacement of a light source from 9.2 meters to 9.5 meters away from a
target, which might be measured in (candela/cm^2)/meter).

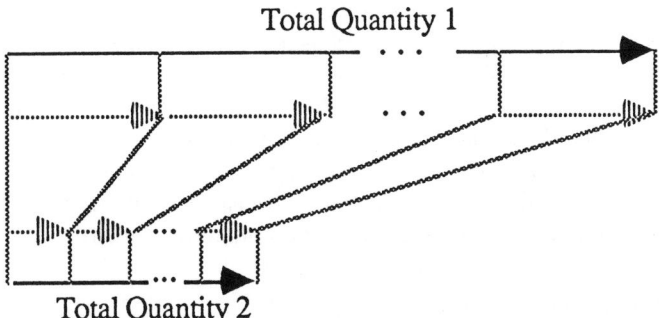

Fig. 2.   An image of rate that entails proportionality between total accumulations in relation to accumulations of accruals. The two quantities vary in relation to each other so that the fractional part of Total Quantity 1 made by any accumulation of accruals or parts thereof within Total Quantity 1 is the same as the fractional part of Total Quantity 2 made by a corresponding accumulation of accruals or parts thereof within Total Quantity 2.

A general scheme for rate entails coordinated images of respective accumulations of accruals in relation to total accumulations. The coordination is such that the student comes to possess a preunderstanding that the fractional part of any accumulation of accruals of one quantity in relation to its total accumulation is the same as the fractional part of its covariant's accumulation of accruals in relation to its total accumulation. More formally, this can be expressed as

$$\frac{accumulated\ accruals\ 1}{total\ accumulation\ 1} = \frac{accumulated\ accruals\ 2}{total\ accumulation\ 2}$$

although expressing it this way does not capture the dynamics of an image of covariation that I am trying to convey. I have tried to capture this image of covariation in constant ratio in Figure 2. Another way to interpret the diagram in Figure 2 is that it is the product of one's coordination of iterable units (Steffe, 1991 and in press).

A significant aspect of mature images of rate is that accruals and accumulations are two sides of a coin. Two quantities which change in measure (accumulate) so that they remain in constant ratio do so through simultaneous accruals which adhere to the ratio; two quantities which change through accruals in constant ratio have total accumulations which themselves adhere to the ratio. A hallmark of a mature image of rate is that accrual necessarily implies accumulation and accumulation necessarily implies accrual.[4]

THE FUNDAMENTAL THEOREM OF CALCULUS

The Fundamental Theorem of Calculus, developed independently by Newton and Leibniz in the late 1600's, provides what Courant called "the root idea of the whole of differential and integral calculus" (Courant, 1937, p. 111). Its creation made possible the algorithmic development of what we know now as the calculus. It also created a cultural necessity for deeper examinations of, and ultimately the resolution of, relationships between conceptions of discrete and continuous magnitudes, whence the formalization of the real number system (Baron, 1969; Boyer, 1959; Wilder, 1981). While the history of the Fundamental Theorem and the developments it fostered are rich and fascinating as topics in their own right, I will focus on ways of thinking that might make it intelligible to individuals reflecting on relationships between derivative and integral. The relationship between derivative and integral is often stated today as follows:

FUNDAMENTAL THEOREM OF CALCULUS

Suppose f is continuous on a closed interval $[a, b]$.

Part I. If the function G is defined by

$$G(x) = \int_a^x f(t)dt$$

for every x in $[a, b]$, then G is an antiderivative of f on $[a, b]$.

Part II. If F is any antiderivative of f on $[a, b]$, then

$$\int_a^b f(x)dx = F(b) - F(a)$$

(Swokowski, 1991, p. 283)

I shall focus on what Swokowski calls *Part I* of the Fundamental Theorem. This says that if some quantity A has a measure t that ranges from a to b, and if some quantity B has a measure $f(t)$ that is conceived as being a function of the measure of A, and if AB is a quantity made multiplicatively from quantities A and B, then as quantity AB accumulates with variations of A (and hence B), the accumulation of quantity AB changes at a rate that is identical with the measure of quantity B at the upper end of AB's accumulation.[5]

The Fundamental Theorem of Calculus – the realization that the accumulation of a quantity and the rate of change of its accumulation are tightly related – is one of the intellectual hallmarks in the development of the calculus. Prior to Newton's and Leibniz' realization of the Fundamental Theorem, what we now call integration was conceived primarily as the determination of a cumulative amount of some quantity, such as arc length, area, volume, or mass; what we now call differentiation was conceived primarily as the determination of angular velocity, tangency, and curvature (Baron, 1969). But these two classes of problems were conceived separately, and each was developed with techniques limited to the type of problem being addressed.

Although both classes of problems are readily seen to be separately capable of inversion, thus, given the area under the curve or the tangent to the curve in terms of abscissa or ordinate, to find the curve, the relation between tangent method and quadrature [area] is not so immediately obvious. The relation between tangent and arc ultimately became one of the most significant links between differential and integral processes and, for this and other reasons, the problem of rectification became crucial in the seventeenth century. The inverse nature of the two classes of problems was approached in terms of a geometric model by Torricelli, Gregory and Barrow but only with Newton did the relation emerge as central and general. (Baron, 1969, p. 4)

The focus on the two classes of problems mentioned by Baron developed as a natural outgrowth of the realization by Apollonius, Oresme, Viète, Descartes, and Fermat that covariation of two magnitudes can be depicted graphically, so that any problem having to do with accumulation could be represented as the determination of an area and that any problem having to do with rate of change could be represented as the determination of tangency (Boyer, 1959). That is, initial development of ideas of the calculus was being done by mathematicians who had a strong preunderstanding that even though they were focusing explicitly on tangents to curves or areas bounded by curves, they were in fact looking for general solutions to any problem of accumulation or change that could be expressed analytically.

Accounts by Baron (1969) and by Boyer (1959) suggest that Newton became aware of the Fundamental Theorem by way of a very definite image of cumulative variation: that accumulations happen by a process of accrual.[6]

Here [in Newton's development of relationships between derivative and integral] we have an expression for area which was arrived at, not through the determination of the sum of infinitesimal areas, nor through equivalent methods which had been employed by Newton's predecessors from Antiphon to Pascal. Instead, it was obtained by a consideration of the momentary increase in the area at the point in question. In other words, whereas previous quadratures had been found by means of the equivalent of the definite integral defined as a limit of a sum, Newton here determined first the rate of change of the area, and then from this found the area itself by what we should now call the indefinite integral [antiderivative] of the function representing the ordinate. It is to be noted, furthermore, that the process which is made fundamental in this proposition is the determination of rates of change. In other words, what we should now call the derivative is taken as the basic idea and the integral is defined in terms of this. (Boyer, 1959, p. 191)

It is worthwhile to mention that Newton envisioned fluxions (rates of change in quantities) and fluents (flowing quantities made by fluxions) as what we would today call functions. This is one reason why his insight was so important. His method was to start with an analytic expression for a function f that gives the rate of change of some quantity and derive an analytic expression for a function F that gives the cumulative amount of that quantity.

What images might have supported Newton's insight? First, Newton was committed to an image of dynamic quantities, in continual flux, instead of to the more common notion of quantities in fixed, indeterminate states (Kaput, in press). Second, as noted by Baron (1969, pp. 263–266), Newton understood motion as being the unifying concept for his methods to determine tangency (rate of change), curvature, arc length, and quadrature (accumulation). Third, he felt quite comfortable thinking of a continuum as being composed of infinitesimals – quantities as small as one pleases which may be discounted when held in comparison to a quantity which is an order of magnitude larger (Boyer, 1959, pp. 198–200).[7]

Here is one way to take these notions in combination so that the Fundamental Theorem is intuitively clear: In a changing, multiplicative quantity, the total accumulation changes at the rate of the accruals of the constitutive quantities.

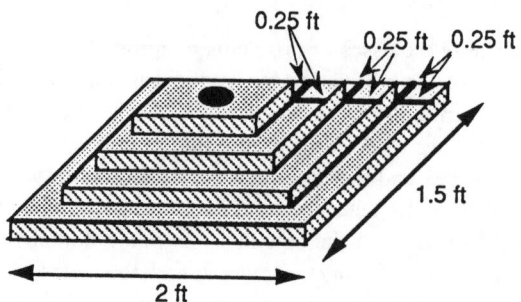

Fig. 3. A mold made of tiers with square bases. Each corner of a tier is offset 0.25 ft perpendicularly from the nearest edge of the tier below it.

For example, suppose you have driven a car for x miles, and that in the next 0.0001 seconds you average 93 km/hr. During that 0.0001 seconds, your *total* driving distance is changing at the rate of 93 km/hr – regardless of how far you have driven. If we imagine that during each infinitesimal period of time you drove at some average speed, and if we could know each of those average speeds, we could reconstruct your total driving distance at each infinitesimal moment of time. Thus, if we were to have an analytic expression which gave us your speed during each infinitesimal period of time, we could, in principle, recover your "distance function." The problem is now one of technique – construct an analytic function whose rate of change differs at most infinitesimally from the rates of change we know you had. This method is not unique to speed and distance, but will apply to any quantity constructed multiplicatively from a rate and another quantity.

A second example will highlight the interrelationships among accumulation of a quantity, accruals in its constituents, and rate of change: Suppose that liquified plastic is being poured into a hollow mold, shown in Figure 3, through a hole in its top. Each corner of a tier is offset 0.25 feet perpendicularly from the nearest edges of the tier below it. Let $v(h)$ represent the volume of plastic in the mold as a function of the plastic's height h from the bottom of the mold. At what rate is $v(h)$ changing with respect to h when the plastic is filling the third tier? The *total* volume is changing at the average rate that the third tier is filling, which is simply the volume of the tier divided by the height of the tier, which in turn is $(A(base) \cdot height)/height$.

These examples bring out two important images: (1) thinking of quantities as being composed multiplicatively of two other quantities, and (2) thinking in terms of infinitesimals.[8] In the first example, increments of distance are conceived of being made by traveling for some small amount of time at some speed. In the second example, increments of volume are conceived as being made by taking some base area to a varying height. In either case, the accumulating quantity is imagined to be made of infinitesimal accruals in the quantities which, composed multiplicatively, make up the accruals in the accumulating quantity. When one

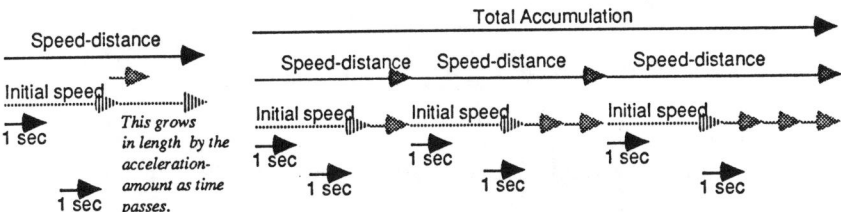

Fig. 4.   Acceleration – the rate at which the speed-distance per time-unit grows.   Image is of acceleration happening in jumps.

of those quantities is the rate at which the quantity changes over an infinitesimal interval, then the total accumulation changes over any infinitesimal interval at the quantity's rate of change over that infinitesimal interval.

EARLY IMAGES OF THE FUNDAMENTAL THEOREM OF CALCULUS

While we cannot expect students to recreate the discoveries of Newton, we can look for kinds of reasoning which would provide us with starting points to develop instructional and curricular approaches oriented at students' development of imagery and forms of expression to support their later insight into important ideas in the calculus. In this section I will report one teaching experiment which attempted to do this. The teaching experiment was with Sue, a seventh-grader, and the content of the teaching experiment were the ideas of speed and acceleration.

An image of acceleration is that "speed grows with time." I have depicted this image in Figure 4. The quantification of acceleration is the determination of by how much the speed-distance grows with each passing unit of time. The complication that acceleration introduces in students' comprehension of situations is not so much in the accrual as in imagining the accumulation.

I depicted the accumulation in Figure 4 as happening only in whole-increments of time. This depiction seems justified not as an accurate portrayal of the most sophisticated understanding of acceleration, but as an intermediate image that becomes refined through the study of limiting processes typically developed in calculus.

Sue's work on a problem having to do with acceleration is presented below. I had already established that Sue possessed the scheme of operations entailed in Figure 2 in the context of a unit on reasoning about speed as a rate (Thompson, in press a).

EXCERPT 1

1.1	Pat:	Imagine this. I'm driving my car at 50 mi/hr. I speed up smoothly to 60 mi/hr, and it takes me one hour to do it. About how far did I go in that hour?
1.2	Sue:	*(Long pause. Begins drawing a number line.)*
1.3	Pat:	What are you doing?

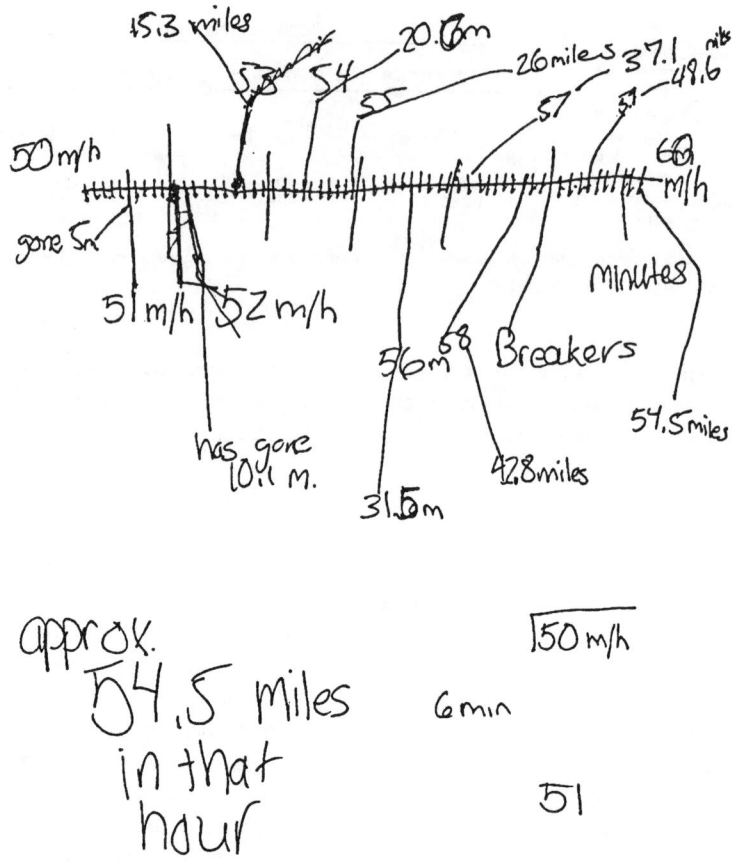

Fig. 5. Sue's scratch work for "How far did I go while I took one hour to speed up from 50 mi/hr to 60 mi/hr?".

1.4	Sue:	I figure that if you speed up 10 miles per hour in one hour, that you speeded up 1 mile per hour every 6 minutes. So I'll figure how far you went in each of those six minutes and then add them up. (*See Figure 5.*)
1.5	Pat:	(*After Sue is finished.*) Is this the exact distance I traveled?
1.6	Sue:	No ... you actually traveled a little farther.
1.7	Pat:	How could you get a more accurate estimate?
1.8	Sue:	(*Pause.*) I could see how far you went every time you speed up a half mile per hour.

Figure 5 shows Sue's work. She assumed that Pat accelerated at the rate of 10 (mi/hr)/hr, which would be equivalent to 1 (mi/hr)/$\frac{1}{10}$ hr. She then assumed Pat drove for one-tenth of an hour (6 minutes) at 50 mi/hr, then one-tenth hour at 51 mi/hr, and so on. She then determined how far Pat would go in each of these one-tenth hour periods.

Sue's solution to estimating the distance I traveled while accelerating has the

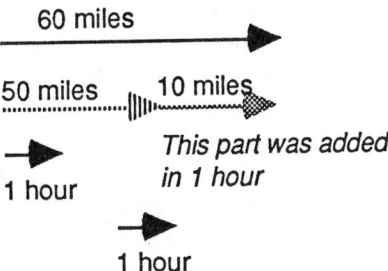

Fig. 6. Sue's image of increasing a car's speed from 50 miles per hour to 60 miles per hour as being the result of increasing the speed-distance by 10 miles at a uniform rate of 1 mile every one-tenth of an hour.

structure of a Riemann sum. It would be expressed formally as

$$\Delta V = final\ speed - initial\ speed\ \text{(a number of miles per hour)}$$

$$\Delta T = final\ time - initial\ time\ \text{(a number of hours)}$$

$$\Delta t = \frac{\Delta T}{\Delta V}\ \text{(a number of hours)}^9$$

$$\Delta v = 1\ mi/hr$$

$$d = \sum_{i=0}^{n-1}(initial\ speed + i\Delta v)\Delta t\ \text{(a number of miles)},$$

which says that you first imagine that the increase in speed is distributed evenly across the number of hours you take to speed up, then pretend that you go at a constant speed within each increment of time and add up how far you go in each of them.

What I wish to draw attention to is Sue's initial inference that Pat's speed increased by one mile per hour every one-tenth of an hour. This seems to be the crucial inference that got her going, and this inference seems to be based on an image of total acceleration like that shown in Figure 6.

Sue's inference was that since 10 mi/hr was added to Pat's speed in one hour, this was the same as adding 1 mile per hour to Pat's speed every one-tenth of an hour.[10] This suggests an image of acceleration that falls between a concept of speed and a concept of continuously accelerated speed.

While we can be inspired by the sophistication of Sue's reasoning, we should take care not to read too much into it. Evidently, Sue had an operational rate scheme, as evidenced by her coordination of acceleration, velocity, and distance, but she had not yet formalized these coordinations so that she could express them analytically. In Excerpt 1, Sue's construction of distance traveled while accelerating for one hour was for a specific increase in speed over a specific amount of time. She was not able to express the general structure of her approach as I did in my summary after Figure 5. To accomplish such a summary, Sue would have needed to encapsulate her method within a language so that her entire process is captured by an expression which describes local behavior of the process.

Another aspect of Sue's reasoning which will be important in the sequel is that her image of the situation seemed not to entail the continuous growth of velocity, and hence of distance, *during* the periods of acceleration. She did not realize that the questions I asked her could have been asked about *any* moment of time during the two respective periods of acceleration, and that her calculational method would, in principle, yield an approximate distance traveled at *each* moment while accelerating. This is not to disparage Sue's reasoning. Rather, it is to point out a significant difference between Newton's and Sue's perspectives. Sue saw completed growth. Newton saw cumulative growth varying immediately as a function of time.

The inspiration we can draw from Sue's example is that there are early forms of imagery which we might draw upon pedagogically in teaching ideas of the calculus. It remains an open question as to how we might provide occasions for students to transform those images into others which are propitious for insight into the calculus.

A TEACHING EXPERIMENT ON THE FUNDAMENTAL THEOREM OF CALCULUS

To study students' insights into the Fundamental Theorem of Calculus I devised a teaching experiment for a group of students enrolled in a course on computers in teaching mathematics. I had two reasons for choosing this group of students. The first was serendipity – this course is structured to have students first experience what it means to conceptualize important ideas in mathematics deeply and then devise instruction to foster the same experiences with their students. A focus on the Fundamental Theorem fits naturally within this structure. The second reason is that I hoped to gain insight into the kinds of understandings and orientations students take with them from introductory calculus and into secondary mathematics classrooms.

The Students

The group was composed of 7 senior mathematics majors, 1 senior elementary education major, 10 masters students in secondary mathematics education, and 1 masters student in applied mathematics. Seventeen students had completed 3 semesters of calculus with grades of B or better, while the other two had grades of C. Seven students had taken advanced calculus and four were currently enrolled in advanced calculus.

In a preliminary assessment only one student, a teacher of Advanced Placement calculus, gave a satisfactory definition of the definite integral of a function; the expression $x^{n+1}/(n+1)$ was the most common response. Only four students gave a satisfactory definition of the derivative of a function; statements about the slope of a tangent were the most common response. In response to the question

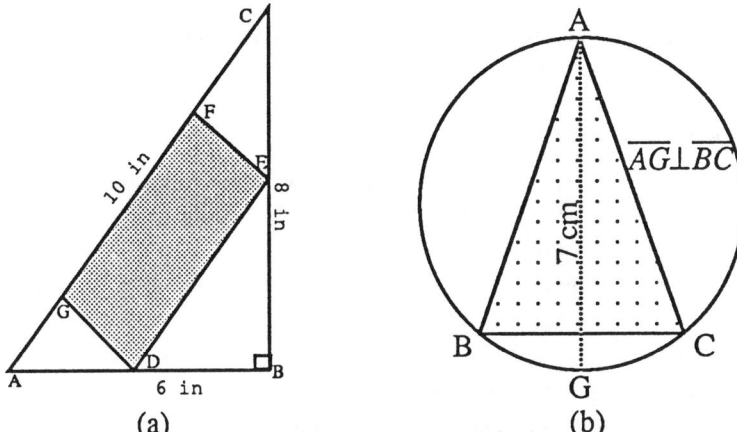

(a)   (b)

Fig. 7. Diagrams accompanying this problem: Use a graping program to find the dimensions of the rectangle in (a) and the triangle in (b) which produce the largest possible area.

"What letter goes in the blank to define this function: $F(\_) = \int_a^x f(t)dt,$ " 16 students said that the letter t goes in place of the blank. Fifteen of 19 students solved a simple optimization problem, and 10 of 19 solved a complex optimization problem. Both optimization problems were taken from a calculus text.

The concept of function was problematic for many students. Six of 19 students could express the area of the rectangle in Figure 7a or the area of the triangle in Figure 7b as a function of some quantity (e.g., area of the rectangle as a function of the length of $\overline{AD}$ in Figure 7a) so that it could be graphed over a suitable domain by a graphing program. A common complaint was that there was not enough information to "solve for the area." Four of 19 students gave satisfactory explanations for why the graph of $f(x) = \cos(8\sin(3x))$, $x\epsilon[-\pi, \pi]$, behaves as it does. Most explanations made no reference to the behavior of $8\sin(3x)$ or to the fact that any function will be periodic if and only if its argument is periodic modulo some modulus.

Classroom conversations and self-reports of students' high school and college mathematical experiences suggested that they and their instructors had engaged largely in "symbol speak" – talking about notations and notational actions without mentioning an interpretation of the notations themselves. As a result, students had learned to focus their attention on internalizing patterns of figural actions – the kinds of things to write, where to write them, and so on. Later excerpts will show the ways in which students expressed an orientation to notational action patterns sans interpretation during the teaching experiment.

The Teaching Unit

The class met twice weekly for 1.5 hours each meeting between February 2 and March 4, 1993, for a total of 10 meetings. Students had ready access to a computer lab or had a computer at home on which to work on assignments. The last two meetings – those in which the Fundamental Theorem of Calculus was discussed – were videotaped and transcribed. A small-group session after the last meeting was also videotaped and transcribed.

The teaching experiment was structured to focus on four phases of conceptual development. These were:

Phase I: Analyze behavior of functions' graphs; explain their behavior; Model situations using functions and derive information about situations from graphs (3 meetings)

Phase II: Average rates of change; functions which give average rates of change over all intervals of a fixed length (2 meetings)

Phase III: Accumulations of change: Riemann sums (2 meetings)

Phase IV: Relationships among variable quantity, accumulation of change, and rate of change of accumulation (2 meetings)

The first phase focused on orienting students to reconstitute their images of function so that it would be based on images of covariation (Thompson, in press b). The second phase focused on having students enrich their notion of average rate so that they could express it as a difference quotient that reflected average rate of change over an increment of some quantity. The third phase focused on having students conceptualize Riemann sums as functions that describe an approximate accumulation of one quantity with respect to variations in another. The unit was intended to culminate in Phase IV by asking students to bring these separate developments together in the context of problems that highlighted the inverse relationship between accumulation and accrual so that they would have an occasion to construct, for themselves, the Fundamental Theorem of Calculus. It was my hope that students would construct the Fundamental Theorem of Calculus; the larger aim of the teaching experiment, however, was to highlight aspects of their conceptions and orientations that might facilitate or obstruct such a construction.

It is important to note that all through the teaching experiment I gave explicit attention to students' images of mathematical activity, with special reference to uses of notation and the construction of explanations. It was essential that they come to interpret notation as someone's attempt to say something – and hence they should reflect on what was intended to be said, and that they should use notation as a medium for expressing their images, inferences, and methods. But to express images, inferences, and methods it was also essential that they come to take these as important activities upon which to reflect. This was an uncommon orientation for most students, and the details of our contract were continually renegotiated.

I will briefly describe the teaching experiment's first three phases to illustrate the nature of instruction and orientation I took to the subject, and to give a sense of the students' orientations. I will describe the fourth phase in detail.

Phase I: Functions, Graphs, and Models (3 meetings)
Students were given two assignments aimed at their developing insight into the behavior of functions by examining the behaviors of their graphs. Examples

TABLE I

Sample tasks from Phase 1 of the teaching experiment. Roman numerals indicate assignment number during the teaching experiment

I.2.	Investigate the behavior of these functions. Explain *why* they behave the way they do. [Note: A good explanation is one which, if understood ahead of time, would have allowed you to *predict* the behavior of the function.]

$$f_2(x) = x\sin(1/x)$$

$$f_3(x) = \cos(x) + 0.01\text{abs}(\cos(100x))$$

$$f_6(x) = x^2 \bmod 2$$

Answer each of the following questions by constructing appropriate functions and then using Analyzer to graph the functions and estimate the question's answer. For each problem, hand in:
- a labelled diagram,
- a statement of what the function represents,
- an explanation of what the function's graph shows you about the situation,
- a note about what you looked for in the graphs to answer the question.

II.1.	Jamie Johnson rides frequently with her father to Chicago. One one particular trip it took 2 hours for them to travel the 110 miles from home to Chicago. They made the trip in two parts. Jamie kept an eye on the speedometer and estimated that in the first part they averaged 40 miles per hour. She estimated that in the second part they averaged 60 miles per hour. About how long did they drive in each part of the trip?
II.7.	Statistical data from trucking companies suggests that the operating cost of a certain truck (excluding driver's wages) is $12 + x/6$ cents per mile when the truck travels at x miles per hour. If the driver earns $6.00 per hour, what is the most economical speed to operate the truck on a 400 mile turnpike where the minimum speed is 40 miles per hour and the maximum speed is 65 miles per hour?

of tasks from these assignments are given in Table I. Classroom discussions emphasized that Cartesian graphs are made of points, and the points in a graph are positioned in a way that reflect each value of a function in relation to the argument that produces that value. Functions as models of dynamic situations were emphasized through problems like II.1 and II.7 (Table I).

Phase II: Average Rates and Functions (2 meetings)

The derivative of a function is typically developed pointwise. That is, it is defined with the understanding that x in the expression

$$f'(x) = \lim_{h \to 0} \frac{f(x+h) - f(x)}{h}$$

is fixed relative to h. My instruction on rates of change drew from an example developed by David Tall (Tall, 1986; Tall et al., 1988) wherein x in the definition of $f'(x)$ is free to vary for each value of h. This alternative approach to the derivative has two very natural interpretations. The first is that for a fixed value of h, the function

$$f_h(x) = \frac{f(x+h) - f(x)}{h}$$

gives the average rate of change of f over every interval of length h contained in the domain of f. The second is that

$$f_h(x) = \frac{f(x+h) - f(x)}{h}$$

gives the slopes of every secant which connects the points $(x, f(x))$ and $(x + h, f(x + h))$. The second interpretation supports an image of a "sliding secant" – slide an interval of length h through the domain of f, thereby sliding the secant defined over that interval, and keep track of the secant's slope. The relationship of either interpretation to the standard definition of the derivative is that as we let h approach 0 we produce a family of functions that converges to the function which gives the instantaneous rate of change of f at every value in the domain of f where the pointwise limit exists.

The reason for my taking this approach to the derivative is that the notion of function is always uppermost in any discussion of a function's rate of change. It also encourages students to think of a function's rate of change in concrete settings in ways that are consistent with ideas of rate of change over some interval. Finally, I intended that their image of a function's average rate of change over a small interval would come into play when thinking of the relationship between accumulations and accruals in Phase IV.

Table II presents sample tasks from Phase II of the teaching experiment. The tasks here were oriented toward conceptualizing the derivative as a function that is approximated by a "Newtonian ratio," a jargon phrase concocted during the teaching experiment to refer to $f_h(x)$.

I was suprised by the nature of students' difficulty in interpreting the functions they defined for III.4 (Table II). Excerpt 3 provides an interchange between myself and two students in the computer lab after they had developed and graphed the function

$$r(x) = \frac{d(x + 0.1) - d(x)}{0.1},$$

where $d(x)$ was defined as $d(x) = 16x^2$. Bob is a high school mathematics teacher, Alice is a mathematics education masters student.

EXCERPT 3

3.1	Bob:	We're having trouble making sense of what we're looking at.
3.2	Alice:	Or even what we did!
3.3	Pat:	Okay, what is this function you typed? What does it represent?

TABLE II

Sample tasks from Phase 2 of the teaching experiment. Roman numerals indicate assignment number during the teaching experiment

---

When an object falls from a resting start, the distance it has fallen t seconds after being released is given by the function $d(t) = 16t^2$ (assuming we ignore resistance).

III.3.   An engine fell off a DC 9. What was the engines average vertical speed between 3.1 seconds and 3.2 seconds after it started falling? Between 3.2 and 3.3 seconds? Between 3.15 and 3.25 seconds? (Answer these questions using just paper-and-pencil.)

III.4.   Use Analyzer to produce a graph of the engine's average vertical speed over *every* 1/10th second interval. (*Don't fall into the trap of thinking that the only 1/10th second intervals are (0, 0.1], (0.1, 0.2], (0.2, 0.3], and so on. In stead, think of a "sliding interval" that has every value in the domain as its left end point.*)

III.5.   Generalize part (III.4) so that your function uses a parameter. Play around with different values of the parameter to generate a family of functions which approximate the function that gives the engine's vertical speed at every *instant* of time after it began falling.

III.8.   Jayne, the clas trouble maker, asked a question about (III.5). She said, "If we think of an object at an *instant* of time, then it didn't move any distance over that instant of time, and it didn't take any time to move nowhere. So, what can it possibly mean to talk about the object's *speed* at an instant of time when speed is about moving some distance in some amount of time?" Comment on Jayne's dilemma.

III.10.   Use the technique developed in [earlier problems] to define a function whose values approximate the instantaneous rate of change of the function $g(x) = \cos(x)e^{\sin(x)}$ at all values of x in $(-10, 10)$.

3.4	Alice:	That's what we can't figure out.
3.5	Pat:	How did you come up with it at all?
3.6	Bob:	We just put letters in for numbers [referring to their solutions to III.3, Table II].
3.7	Pat:	Okay, let's take it a piece at a time. What does $d(x + 0.1)$ represent?
3.8	Bob:	How far it went in one tenth of a second.
3.9	Alice:	How fast it is going.
3.10	Pat:	Well ... I don't understand how you came up with your interpretations.
3.11	Alice:	I was guessing (*laughs*).
3.12	Bob:	It's like this ... $d(x)$ gives how far the engine dropped in x seconds, so $x + 0.1$ is another tenth of a second. So $d(x + 0.1)$ gives how far it went in that extra tenth.
3.13	Pat:	How far it went in just that tenth of a second, or how far it fell altogether in $x + 0.1$ seconds?
3.14	Alice:	Oh ... it has to be how far it fell in the whole amount of time.
3.15	Bob:	I don't ... (*pause*)
3.16	Pat:	Let's change the subject for a little while. If I were to tell you how far this engine fell in 7 seconds, what would you need to know to tell me how far it fell in the last two seconds?
3.17	Bob:	(*Pause.*) How far it fell in the first 5 seconds.

3.18	Alice:	Then you'd subtract.
3.19	Pat:	You'd subtract what to get what?
3.20	Alice:	Subtract how far it went in 5 seconds from how far it went in 7 seconds to get how far it fell in the last 2 seconds.
3.21	Pat:	How would you calculate the engine's average speed during those last 2 seconds?
3.22	Bob:	Divide by 2.
3.23	Pat:	Divide what by 2?
3.24	Both:	The distance it went in the last 2 seconds.
3.25	Pat:	Now, tell me again what $d(x + 0.1)$ and $d(x)$ represent?
3.26	Bob:	How far ... how far ...
3.27	Alice:	How far it fell in $x + 0.1$ seconds and how far it fell in x seconds.
3.28	Pat:	Okay, what does the difference of those two represent?
3.29	Bob:	How far it fell in the last tenth of a second?
3.30	Pat:	Not necessarily the last tenth, just the tenth of a second after x seconds of falling. (*Pause.*) Now, what does $r(x)$ represent?
3.31	Both:	How fast it went during that tenth of a second.
3.32	Pat:	Was it always going one speed during that tenth of a second? (*Long pause.*)
3.33	Alice:	Oh ... its *average* speed during that tenth of a second!
3.34	Pat:	Okay! Now, what does the graph of $r(x)$ represent?
3.35	Alice:	How fast ... how fast ...
3.36	Bob:	It's average speed ... after ... (*to himself*) when?
3.37	Alice:	It's average speed ... over ... over ... every one-tenth interval ... one-tenth second. Over every one-tenth second interval of time!
3.38	Bob:	Oh.
3.39	Pat:	Okay (*enters "$r(1.5)$" at keyboard; program prints "49.60"*), this says that $r(1.5)$ is 49.6. What does that mean?
3.40	Bob:	It was going 49.6 feet per second after one and a half seconds.
3.41	Alice:	It went an average speed of 49.6 feet per second when it fell from 1.5 seconds to 1.6 seconds.

Bob's difficulty was not uncommon. Those students who experienced difficulty seemed to want to think of the difference quotient as "the derivative" and interpret it as "how fast it [the function] is changing," without interpreting the details of the expression as an amount of change in one quantity in relation to a change in another. Several students chose to write the difference quotient in their homework as

$$\frac{f(x + h) - f(x)}{(x + h) - x}.$$

I presume this was a mnemonic to help them keep in mind that the denominator was also a difference and that the quotient evaluated a multiplicative comparison of changes.

My intention for Item III.8 (Table II) was to orient students to thinking of instantaneous velocity as a limit of average velocities. In fact, students' responses surprised me. Of the 12 responses turned in, all said essentially that they would explain to Sue that "instant" was not really an instant, but an amount of time so small that it was virtually indistinguishable from zero seconds.

Phase III: Riemann Sums (3 meetings)

I introduced Phase III with a discussion of Sue's problem and solution, as presented earlier in this article (Figure 5). I was struck by the direction taken by students: A consensus emerged that, had they been Sue's teacher, they would have had Sue "discover" that she could just multiply the amount of time taken to speed up by the mean of the beginning and ending speeds. Sue's solution method was, to them, a rather clumsy way to approximate "the correct answer." I asked, "Does Sue's solution have anything to do with calculus?" "No." I then presented Sue's problem with a variable acceleration, asking "Which method will generalize to this new setting – yours or Sue's?" Eventually they demurred that there might have been more sophistication in Sue's reasoning than they originally recognized.

Instruction during Phase III focused on conceptualizing a Riemann sum as a function and on conceptualizing dynamic situations as representable by Riemann sums (see Table III). A major difficulty for many students was to express functional relationships in situations analytically, and to coordinate their images of functional covariation of two quantities with an image of accumulation by way of accruing "chunks" of a quantity. The notion of a Riemann sum as presented in Phase III – an approximation to a variable accumulation – often conflicted with their images of definite integral and Riemann sum as applying only in situations involving *fixed* amounts of some quantity (the typical scenario in freshman calculus). This conflict revealed itself in a number of ways – a common one being that a student would write an expression for a Riemann sum, but with an image that what he or she was finding was a *total* amount of a quantity (e.g., total work, area, volume, etc.) instead of a varying amount of the quantity.

I gave special attention to items like IV.5 and IV.6 in Table III. The reason for this was to give students an occasion to reflect on the details by which the process of Riemannian summation assigns values to its argument. The first case (IV.5) corresponds to assigning a fixed subinterval length in any partition of the interval $[0, x]$. For $x \epsilon [i\Delta x, (i+1)\Delta x)$ the expression $[x/\Delta x]$ is constant, so

$$\sum_{i=1}^{x/\Delta x} f(i\Delta x)\Delta x$$

is constant over that interval,[11] and hence

$$\sum_{i=1}^{x/\Delta x} f(i\Delta x)\Delta x$$

produces a constant function over each of the subintervals through which x varies – a step function. The second case (IV.6) corresponds to assigning a fixed number of subintervals in any partition of $[0, x]$. As x varies, the number of subintervals

TABLE III

Sample tasks from Phase 3 of the teaching experiment. Roman numerals indicate assignment number during the teaching experiment

IV.1.	a.	Use Analyzer and Riemann sums to produce a graph of the approximate velocity of a car during its first 10 seconds of accelerating from a standing start when it accelerates at the rate of 11.5 mi/hr/sec.
	b.	Use Analyzer and Riemann sums to produce a graph of the approximate distance covered by a car during its first 10 seconds of accelerating from a standing start when it accelerates at the rate of 11.5 mi/hr/sec.
IV.2.		Use Analyzer and Riemann sums to produce a graph of the volume of water in a conical storage tank that is 25 feet high and 30 feet wide at the top. Express the volume as a function of the height of the water above the tip of the cone.
IV.5.	a.	How might you think of the expression $$\sum_{i=1}^{x/\Delta x} \cos(i\Delta x)\Delta x$$ to understand that it defines a Riemann sum evaluated at every value of x in your domain?
	b.	Explain why the Riemann sum defined this way *always* produces a step function, regardless of the value of Δx (assuming it is not zero).
IV.6.	a.	How might you think of the expression $$\sum_{i=1}^{n} \cos\left(i\frac{x}{n}\right)\frac{x}{n}$$ to understand that it defines a Riemann sum evaluated at every value of x in your domain?
	b.	Explain why the Riemann sum defined this way *never* produce a step function, regardless of the value of n.

in the partition remains the same, but the subintervals "stretch" proportionally as x gets proportionally larger. In both cases I emphasized that they should try to imagine the process of Riemann summation as happening so rapidly that they could think of x varying freely and the process would keep up with it. That is, as x varies, the process of summation happens at each value of x, and the process happens "so rapidly that it doesn't slow x down – you can slide x along its domain and not feel any resistance from the process as it tries to keep up."

Phase IV: The Fundamental Theorem of Calculus (2 meetings)

I did not introduce the Fundamental Theorem of Calculus as such. Instead, I continued a discussion of one Riemann sum problem that students had found particularly troublesome. The problem was:

Hexane is a gas used for industrial purposes. Clentice Smith of Cargill Corp., Bloomington, IL in November, 1989 requested a graph that will give the approximate volume of hexane (measured in cubic inches) held by the tank shown in Figure 8. Use Analyzer and Riemann sums to produce

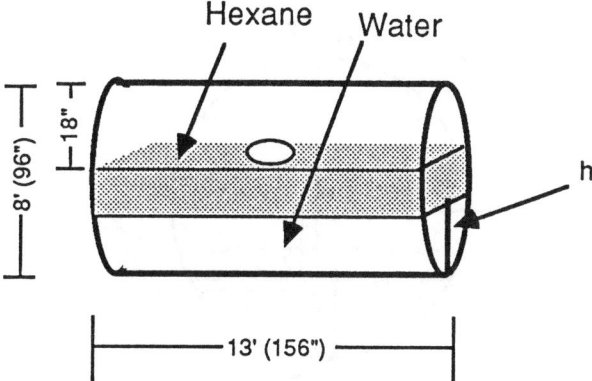

Fig. 8.   A tank with water in the bottom and hexane sitting atop the water. The hexane always reaches the hole in the tank's side regardless of the height of the water.

such a graph. Express the volume of hexane as a function of the height of the water (measured in inches).

Assumptions

- The face of the tank is a disk (i.e., a region bounded by a circle)
- the shape of the tank is cylindrical
- the hexane sits atop the water
- the dimensions of the tank are as shown
- a hole in the tank resides $18''$ vertically from the top of the tank
- the hexane always reaches the bottom edge of the hole.

My intention with this problem was to recap students' solutions and use the discussion as a setting for asking about, what would turn out to be, the Fundamental Theorem of Calculus. I intended to do this by graphing the width of a horizontal slice of the tank's face as a function of the slice's height from the floor, graphing the area of the face's water-covered portion as a function of the water's height, and then ask about how fast the water-covered portion's area changes with respect to the water's height. I presumed that students would suggest a difference quotient to estimate the function which gives rate of change of area as a function of the water's height, and I planned then to graph this difference quotient and ask students to compare its graph with the graph of the horizontal-slice width as a function of height (anticipating that the two graphs will appear to be identical). The culminating question would be, "If this graph is of the width of a horizontal slice as a function of its height from the tank's bottom, and the other graph is of the rate of change of area as a function of the region's height from the tank's bottom, then why do they look the same? Is there some reason for it, or is it just coincidence?"

The session began with one student's, Blake's, presentation of his solution to the problem. He established that the main aspect of this problem was to express the area of the water-covered region of the tank's face as a function of the water's height from the bottom. Blake then defined a function to give the width of an

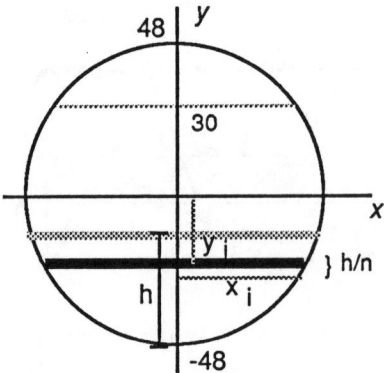

Fig. 9. Face of cylindrical tank. Water level is at height h measured from tank's bottom, with ith rectangle in partition highlighted; x_i and y_i are coordinates of the ith rectangular piece's lower-right corner.

arbitrary chord as a function of its height (Figure 9) and set up an appropriate Riemann sum as a function of the water's height.

After Blake had completed his presentation I displayed his equations on a projector screen.[12]. They are presented below as Equation Set 1. The function $x(h)$ gives the x-coordinate of a chord's right endpoint expressed as a function of the chord's height above the bottom of the circular face (Figure 9). The function $w(h)$ gives the chord's width. The function $A(h)$ gives the approximate area of the water-covered portion of the tank's face as a function of water's height. The function $V(h)$ gives the approximate volume of the hexane as a function of the water's height.

$$n = 20 \hspace{4cm} \text{E1.1}$$

$$x(\mathbf{h}) = \sqrt{48^2 - (-48 + \mathbf{h})^2} \hspace{3cm} \text{E1.2}$$

$$w(\mathbf{h}) = 2x(\mathbf{h}) \hspace{4cm} \text{E1.3}$$

$$z = w(h) \hspace{4cm} \text{E1.4}$$

$$A(\mathbf{h}) = \sum_{j=1}^{n} w\left(j\frac{\mathbf{h}}{n}\right)\frac{\mathbf{h}}{n} \hspace{3cm} \text{E1.5}$$

$$y = A(h) \hspace{4cm} \text{E1.6}$$

$$\text{Total Area} = A(78) \hspace{3cm} \text{E1.7}$$

$$V(\mathbf{h}) = 156(\text{Total Area} - A[\mathbf{h}]) \hspace{2cm} \text{E1.8}$$

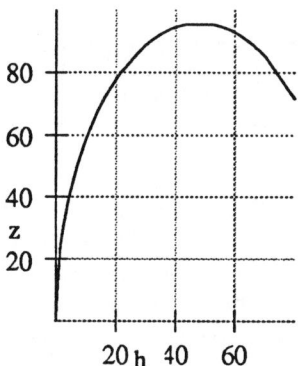

Fig. 10. Graph of $z = w(h)$, the width of a cross section as a function of the cross section's height above the cylinder's bottom.

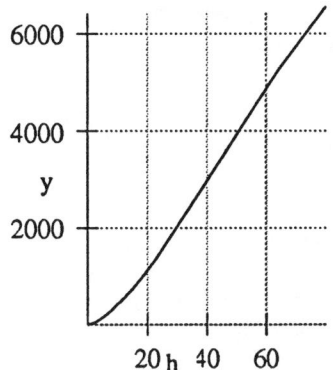

Fig. 11. Graph of $y = A(h)$, the approximate area of the tank face's water-covered portion as height of water increases.

$$v = V(h) \hspace{10em} \text{E1.9}$$

Equation Set 1: Blake's system of equations and functions for the Hexane problem.

We discussed Figure 9 and its relationship to the functions $x(h)$ and $w(h)$, shown above as E1.2 and E1.3, and we discussed the graph of $w(h)$ [Figure 10].

The discussion of $A(h)$ first centered around interpreting its construction, which was not straightforward for some who still had questions. After I was satisfied that everyone understood the construction of $A(h)$ and $V(h)$, I displayed a graph of $y = A(h)$ (Figure 11) and then redirected the focus of the lesson by asking about the rate of change of area of the face's water-covered portion as the water's height increases. The ensuing discussion is given in Excerpt 4.

EXCERPT 4

4.1	Pat:	Let me back up a little [*scrolls back to equation E1.5*]. I want to ask you a question. We had this area function [*highlights E1.5*]. Suppose that I ask you the question, "How could we get a function that approximates how fast the area is changing as the height increases?" [*moves hand upward to indicate an increasing water height.*]
4.2		*Long pause.*
4.3	Student:	Tangent to the slope of the line.
4.4	Bob:	You basically take a tangent at any point on the curve ... on your area function.
4.5	Pat:	Okay. Do you know how to do that?
4.6	Bob:	*Pause.* Basically, by taking limits ... I'm trying to remember this stuff.
4.7	Jim:	Isn't it just that limiting thing that we've been doing?
4.8	Pat:	That limiting thing?
4.9		*Laughter.*
4.10	Alf:	It's just the difference quotient, isn't it?
4.11	Pat:	Alf?
4.12	Alf:	The difference quotient, where
4.13	Jim:	The moving secant line.
4.14	Alf:	Yeah.
4.15	Alice:	Oh yeah!
4.16	Alf:	It would be f of x plus h minus f of x all over h.
4.17	Pat:	What would that give you?
4.18	Student:	The speed.
4.19	Pat:	The speed of what?
4.20	Alf:	The speed for how fast the area is changing.
4.21	Pat:	How does this give you what you say?
4.22	Jane:	It's like average speed.

My question about how fast the area changes with respect to height appeared to take them by surprise. The first two responses seemed to emanate from a concept image of derivative as slope of a tangent. Only when Alf spoke of the difference quotient (¶ 4.10) did the idea of average rate of change over a small interval of height emerge.

How to express the difference quotient of change in area in relation to change in height was problematic for a number of students, despite Alf's suggestion (¶ 4.16). After a consensus emerged on how to express the difference quotient, I entered the equations shown in Equation Set 2 and displayed a graph of the approximate rate of change in area with respect to height (Figure 12).

$$dA(\mathbf{h}) = \frac{A(\mathbf{h} + \Delta h) - A(\mathbf{h})}{\Delta h} \qquad\qquad \text{E2.1}$$

$$\Delta h = 0.01 \qquad\qquad \text{E2.2}$$

$$q = dA(h) \qquad\qquad \text{E2.3}$$

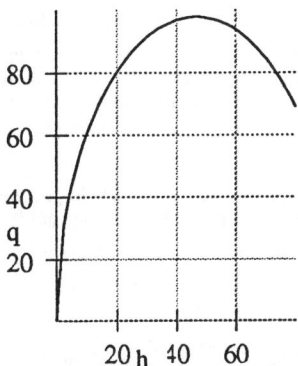

Fig. 12. Graph of $q = \mathrm{d}A(h)$, the approximate rate of change of the water-covered portion's area as a function of the water's height.

Equation Set 2: Equations to define and graph the difference quotient which gives the approximate rate of change of the water-covered portion's area with respect to the water's height.

Excerpt 5, below, presents the discussion immediately following my presentation of Figure 12. It began with "what does this graph show us," but quickly moved to why it appears to be the same as the graph shown in Figure 10.

EXCERPT 5

5.1	Pat:	What does this graph [*Figure 12*] show us?
5.2	Several:	The rate at which the area is changing.
5.3	Pat:	What shows me the rate at which the area is changing when the height is 20 inches.
5.4	Alice:	It's whatever q is when h is 20.
5.5	Pat:	Does this graph look familiar?
5.6	Jane:	It looks like the ... the uh ... the base of the rectangle that we had.
5.7	Roy:	Oh ... of course.
		......
5.8	Pat:	*Moves the graphs shown in Figure 12 and Figure 10 so that they are side by side on the projector screen.*
		......
5.9	Bob:	What was the one on the left again [*Figure 10*]?
		......
5.10	Alf:	The derivative of ... [*several students speak at once*] ...
5.11	Pat:	That's how fast the area is changing as a function of h.
5.12	Bob:	And it's changing in the same respect as that thing [*pointing to diagram shown in Figure 9*] is getting wider.
5.13	Alice:	[*to herself*] That makes sense.
5.14	Pat:	*Pause.* Why?
5.15	Bob:	*Pause.* Why? Because that's what you multiplied it by!

5.16	Pat:	What do you mean, "That's what I multiplied it by?"
5.17	Bob:	You're taking the change in x [*spreading his hand apart horizontally*] and multiplying it by as it changes here [*holding his thumb and forefinger slightly apart vertically*] ... as the chord length changes ... the change in x gives you one of those little rectangular boxes we've been talking about.
5.18	Pat:	Yeah.
5.19	Bob:	Now, as that changes, as it gets larger, then the area is going to get larger. Now, I know ... I got it in my mind but it's not coming out my mouth.
5.20	Pat:	Can anyone reinterpret what Bob is saying? *Pause.*
5.21	Alice:	I'm having the same problem ... how to say it.

Bob's remarks (beginning in ¶ 5.12) seem to have emanated from a loosely articulated collection of images. It appears that he had an image of a chord moving up (¶'s 5.12, 5.17, 5.19), getting wider as it moves up (¶ 5.19). Bob referred to a "change in x" (¶ 5.17), but it seems more like he had in mind what might more appropriately be called "a changed x", "a changing x," "another x," or even perhaps "a bigger x" – where "x" referred to either a chord or the length of a chord. In (¶ 5.17), Bob referred to getting "one of those little rectangular boxes." In (¶ 5.19) Bob referred to the area changing as "that" changes, presumably meaning that the area changes by moving the chord upward, thereby accumulating another "little rectangular box."

Bob's image, as described in the previous paragraph, gives him insight into the accumulation of area, but it is not an image of the rate of change of area with respect to height. Bob still needed to relate the change in area to the change in height – in the same way that one would relate a change in distance to a change in time to develop insight into speed as rate of change of distance with respect to time.

Bob quit his attempt to explain what he had in mind. Alf and Alice then entered the discussion. Alice eventually hypothesized that the two were somehow identical because they were both changing because of being functions of the height.

EXCERPT 6

6.1	Alf:	Isn't it that the area function is the change ...
6.2	Alice:	The area function changes the ...
6.3	Alf:	The area function is actually the change ... or the rate of change ... for the ... [*spreads hands apart*]
6.4	Bob:	What's staying the same in both?
6.5	Alice:	When you change the height, you change the area, and when you change the height the width changes also ... so therefore ... did you follow that?
6.6	Pat:	Go ahead.
6.7	Alice:	So therefore ... when you want to find the rate of change of the area that's going to go along with the rate of change of the width ... since they're both a function of the height, they're going to change the same ... together.
6.8	Pat:	Paul, did you follow what Alice was saying?

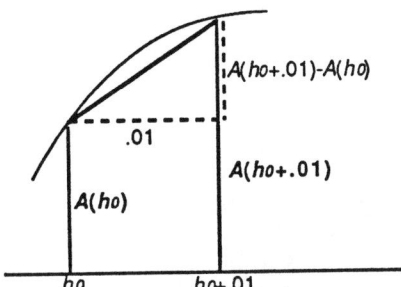

Fig. 13. Increase in area in relation to increase in h of 0.01 inches.

| 6.9 | Paul: | I doooon't knoooow. |
| | | *Laughter* |

Alice's hypothesis regarding the source of similarity between the two graphs (¶ 6.5, 6.7) led eventually to rampant confusion. Students began to misinterpret graphs (e.g., saying that the graph of $z = w(h)$ shows the rate of change of the width, or that the graph of $q = dA(h)$ shows the area as a function of height) and to confuse volume with area. I decided to redirect the discussion to try to emphasize rate of change.

EXCERPT 7

7.1	Pat:	Perhaps it would be helpful to come back to the area function [*points at E1.5*] and its graph [*draws a section of the graph of $y = A(h)$ on the blackboard*]. In terms of the graph, what we're doing at each value of h is to find the slope of a secant over an interval of length 0.01 [*see Figure 13*]. Let's label this point h_0 and this point $h_0 + 0.01$. What is this value [*indicates vertical segment at h_0; note that Figure 13 shows all labels, but the vertical magnitudes were not yet labeled during this exchange*]?
7.2	Bob:	V of h.
7.3	Pat:	Actually, it's A of h_0 – this is the graph of area as a function of height. What is this value [*indicates right vertical segment*]?
7.4	Several:	A of h naught plus point zero one.
7.5	Pat:	Okay. And this [*indicates excess of $A(h_0 + 0.01)$ over $A(h_0)$ in Figure 13*] is the difference between the two ... $A(h_0 + 0.01) - A(h_0)$. [*Writes expression on blackboard. Diagram on blackboard now matches Figure 13*].
7.6		We're looking at a little bit of area ... on the surface of that disk. So here is, if you like, $A(h_0 + 0.01)$ [*draws diagram with \\\\ hash marks; see Figure 14*] and here is $A(h_0)$ [*drawn //// hash marks: see Figure 14*]. When you subtract $A(h_0)$ away[*sweeps hand across //// hashed region*], you're left with $A(h_0 + 0.01) - A(h_0)$. What is this [*sweeps hand across difference region*]? [*Pause.*] This is approximately the width at h_0 times the height of this little piece. The height is approximately what?
7.7	Student:	Delta h.

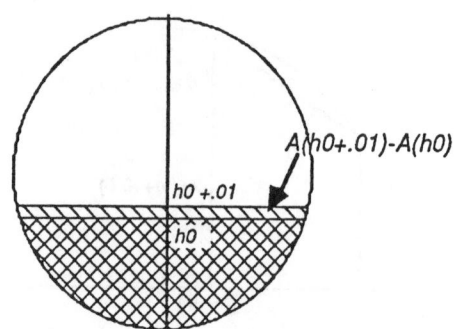

Fig. 14. "A little bit of area." .

$$\underline{dA}(\underline{h_0}) = \overbrace{\frac{A(h_0+.01) - A(h_0)}{\Delta h}}^{w(h_0)\,\Delta h} \approx \frac{w(h_0)\,\cancel{\Delta h}}{\cancel{\Delta h}}$$

Fig. 15. Expression for approximating the average rate of change of region's area over the interval $[h_0,\ h_0 + \Delta h]$.

7.8	Pat:	Point zero one, but in principle your right, it's delta h. [*Writes $w(h_0)\Delta h$ next to difference region.*] So this [*puts a bracket above $A(h_0 + 0.01) - A(h_0)$*] is approximately the width at h_0 times delta h [*writes $w(h_0) \times \Delta h$ above bracket*].
7.9	Alf:	Divide that by delta h.
7.10	Pat:	Yeah ... [*writes fraction bar under $A(h_0 + 0.01) - A(h_0)$, then Δh under fraction bar*] divide that by delta h, and guess what?
7.11	Alf:	
	Bob:	You get the width.
7.12	Pat:	You get approximately the width at h_0. [*See Figure 15.*]
7.13	Jane:	Hmmm.

My presentation in Excerpt 7 was too didactic to glean anything now about how students understood the role of rate in linking $w(h)$ and $dA(h)$. Also, in retrospect, I can see that the idea of rate of change moved to the background, becoming implicit in my remarks. I will return to this point later, in my discussion of the teaching experiment.

The next problem, IV.2 in Table III, asked for a Riemann sum that gives the approximate volume of water in a conical storage tank as a function of the water's height. Students had worked this problem earlier with little difficulty. The functions produced in that solution were $A(h) = \pi(\frac{15}{25}h)^2$, which gives the area of a cross-sectional disk as a function of the disk's height from the bottom of the cone, and $V(h) = \sum_{j=1}^{n} A(jh/n)\,(h/n)$, which gives the approximate volume of water when its height is h. I graphed $y = A(h)$ and $z = V(h)$, and then asked, as before, how we could express the approximate rate of change of the

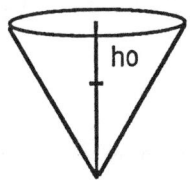

Fig. 16.   Initial diagram in discussion with Sally.

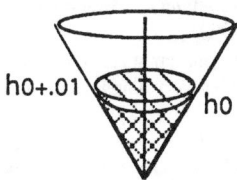

Fig. 17.   Diagram of difference between volume at height $h_0 + 0.01$ and volume at height h_0.

water's volume as a function of its height. Several students suggested graphing the function.

$$DV(h) = \frac{V(h + \Delta h) - V(h)}{\Delta h}.$$

The graph of $DV(h)$ appeared identical to the graph of $A(h)$, and the discussion moved to trying to understand why we should expect them to be the same.

Many of the confusions seen in discussions of the previous problem surfaced again. Students confused "changing" with "rate of change," and confused amount and change in amount. One student, Sally, eventually suggested that we "use the same argument as the last one."

EXCERPT 8

8.1	Pat:	Go ahead and say more.
8.2	Sally:	With ... except for now we would have $\dfrac{V(h_0 + 0.01) - V(h_0)}{0.01}$.
8.3	Pat:	So ... what kind of diagram should we use?
8.4	Sally:	Visualize ... oh.
8.5	Pat:	Here's the cone [*Figure 16*]. How should I shade in $V(h_0)$?

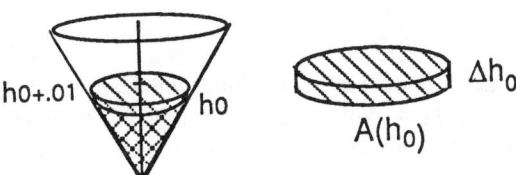

Fig. 18.   Difference of $V(h_0 + \Delta h)$ and $V(h_0)$ approximated by product of area of base times height of cylinder.

8.6	Sally:	So it would be the disk there with Δh as its height?
8.7	Pat:	Just $V(h_0)$.
8.8	Sally:	The disk ...
8.9	Pat:	*Draws a disk centered at h_0.* You said a disk. Now, what is $V(h_0)$?
8.10	Sally:	*Pause.*
8.11	Pat:	That's the volume of the cone from the bottom up to h_0. [*Sweeps hand upward over diagram.*]
8.12	Sally:	Oh ... okay.
8.13	Pat:	So that's everything ... everything up to h_0. [*Shades diagram. Writes* "$V(h_0) = \backslash\backslash\backslash\backslash$"]
8.14	Sally:	So ... V of h_0 plus point zero one or something would be just a little bit above it and would be everything underneath that?
8.15	Pat:	All right. [*Writes* "$V(h_0+0.01) = ////$"] So I would go up [*marks h_0 +0.01*] ... this is $h_0 + 0.01$ [*draws disk at that height; shades in region below disk; see Figure 17*] and take everything under that. And what would we get?
8.16	Sally:	And so when you subtract them you would get ... just a little chunk ... of volume.
8.17	Pat:	That's right. We would get just a little bit of volume [*draws inset of difference; see Figure 18*]. How high is this chunk?
8.18	Sally:	0.01.
8.19	Pat:	0.01. Or ... let's use Δh.
8.20	Sally:	Yeah, Δh.
8.21	Pat:	So, it's Δh high. *Pause.* And this is ... [*indicates base of inset*]?
8.22	Alf:	
8.23	Jim:	The area.
8.24	Pat:	$A(h_0)$, isn't it? So, $V(h_0 + \Delta h) - V(h_0)$ is ... the volume of this little chunk. So, how could we express that given what we know over here [*points to diagram shown in Figure 18*]? Jim?
8.25	Jim:	$A(h)$ times Δh.
8.26	Pat:	[*Completes previously started sentence. Writes $V(h_0 + \Delta h) - V(h_0) \approx A(h_0)\Delta h$.*] So, what happens when we divide by Δh?
8.27	Several:	You just get the area.

As in the discussion of the previous problem, I allowed the idea of rate of change to move to the background, becoming implicit in my remarks. It is not clear from Excerpt 8 whether students understood that the expression

$$\frac{V(h_0 + \Delta h) - V(h_0)}{\Delta h}$$

evaluated an *average* rate of change of volume with respect to height of a cylinder over the interval $[h_0, h_0 + \Delta h]$, which in turn gave the average rate of change with respect to height of the total volume over that interval. We will see that it is unlikely that they understood the discussion to be about rates.

I met a small group of students after class – the number varied during the meeting, starting with four and ending with seven. The purpose of the meeting

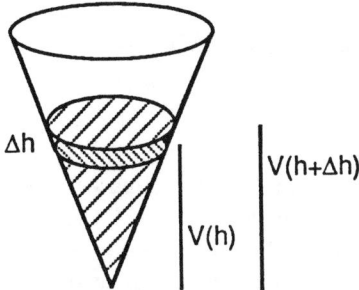

Fig. 19. Students' diagram for identifying amount of change in volume as water rises in height within a conical storage tank.

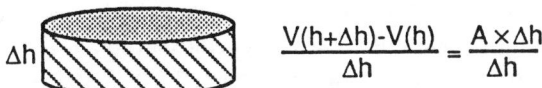

$$\frac{V(h+\Delta h)-V(h)}{\Delta h} = \frac{A \times \Delta h}{\Delta h}$$

Fig. 20. Students' reproduction of diagram drawn during class discussion.

was for students to ask questions and discuss their confusions. Blake, Roy, Adam, and Fred had already begun discussing the "cone" problem, and had drawn Figure 19 and Figure 20 on the blackboard before I joined them.

The four of them had been discussing "how the surface area and volume are related." Blake spoke to me as I joined the group.

EXCERPT 9

9.1 Blake: I'm not getting the connection I mean, in my mind I understand how they're the same. It's kind of like, to put it in words, how the surface area is related to the volume. How ... see, we're talking about the surface area ... the graph of that is the same as the rate of change of the volume ... they're both changing because they're both functions of ... delta h, the change in h.

......

9.2 Blake: To me, it's almost the same thing as when Alf was talking about when you take the derivative of something and then you want to go backwards to it. But they're both ... it's ... it's like I know what I'm thinking, but I can't say it. [*Long pause.*] Uh ... like how are they related? They're related because they're both functions of the height ...

9.3 Pat: Yeah, they're both functions of the height.

9.4 Blake: [*Places thumb and forefinger over cylinder on blackboard; see Figure 20.*] ... of how this disk is changing, I guess [*spreads thumb and forefinger apart*]. As the disk changes, because ... this height ... this delta h always stays the same [*indicates thickness of cylinder with thumb and forefinger*] but the surface area is changing [*sweeps finger in a circular motion over the top of the cylinder*].

9.5 *Long pause. Paul joins the group.*

9.6 I guess it's stupid to say that it's just common sense. As that surface area gets bigger [*moves hands and fingers to show a "growing circle"*] so that in one graph you're looking at the surface area ... the surface area ... that's ... there's no other way that the volume can change. [*Pause.*]

9.7	Pat:	Is except by ...?
9.8	Blake:	As a function of the area. [*Pause.*]
9.9	Pat:	As a function of the surface area?
9.10	Blake:	Yeah, the surface area.

Blake's remarks suggest he was struggling with two sources of meaning for the identity between a cross-section's area and the rate of change of volume with respect to height. In (¶ 9.2) Blake referred to a remark made by Alf, during class, regarding a connection between the derivative of a distance function and the antiderivative of a speed function. This suggests Blake remembered something about an integral of a derivative somehow returning you to an original function. On the other hand, in (¶'s 9.1, 9.4–9.10), Blake appeared to be thinking of a circular disk moving upward, so that the surface area of the disk becomes larger while at the same time water fills the space generated by moving the disk upward. This image resembled Bob's remarks (Excerpt 5, ¶'s 5.17–5.19) and Alice's remarks (Excerpt 6, ¶'s 6.5–6.7) during class about the water-covered area of a tank's face changing as a chord gets wider. Adam's and Fred's comments in the ensuing conversation (Excerpt 10, below) follow Blake's predominant direction of thought – that the two graphs are identical because the two quantities are changing simultaneously.

EXCERPT 10

Discussion continues from Excerpt 9.

10.1	Pat:	Okay. [*Pause. Speaks to Blake.*] So the volume changes ... here's where I'm not clear on what you're saying. It sounds like you're saying that volume changes as the surface area changes.

Alf joins the group.

10.2	Adam:	[*Walks up to the board and points at the top cross section in Figure 19. Turns to Pat.*] Are you just thinking about this as a slab floating on top of the water ... as this [*the slab*] goes up [*moves hand upward*] ... as the surface area gets bigger [*moves hands apart to indicate a growing circle*] ... the volume underneath [*sweeps hand across region below the slab*] is going to change ... the same ... type of rate [*moves hand up and down in front of diagram in Figure 19*]. What I ... I just don't know how to explain it.
10.3	Pat:	Same type of rate?
10.4	Adam:	Well, it's ... this [*slab*] is changing ... is getting bigger [*shows growing circle with hands and fingers*] as you're going up, and this [*volume under slab*] is getting bigger as you're going up [*moves hands as if pushing the slab upward*].
10.5	Pat:	Okay ... so they're both getting bigger.
10.6	Fred:	But ... why is it that they're both the same?
10.7	Pat:	Yes, that's the key question. Why is it that area turns out to be exactly the same as the rate of change of the volume? [*Pause.*] There's a qualitative similarity in that, yes, they are both getting bigger. But Fred asked the key question, "Why is it that they're identical?"
10.8	Fred:	I don't know [*laughter*]. [*Very long pause.*]
10.9	Blake:	Is it so simple that we're just overlooking it, or is it really that hard?
10.10	Pat:	Well, it's partly in front of you.

10.11   Fred: I can see it algebraically when you put it in this kind of form [*Figure 20*], but
                     I guess I have trouble visualizing it.

Adam's remarks in (¶'s 10.2–10.4) are telling in two ways. First, he appears
to have, like Blake and others, an image of a circular disk moving upward, thereby
increasing its area, while the generated space increases in volume. Second, he
seems to have identified "rate" with "change," so that he ended up saying things
like "as the surface area gets bigger ... the volume underneath is going to change
... the same ... type of rate" (¶ 10.2). If Adam was indeed thinking of a rate,
then it was the rate of change of volume with respect to area of the circular cross
section. It appears, however, that by "same type of rate" he meant that the two
quantities change simultaneously and in the same direction (increase) instead of
as an amount of change in one quantity in relation to an amount of change in the
other.

Fred's comments (¶'s 10.6–10.11) are also telling in several ways. First, he
appears not to have an articulated image of "they" in "But ... why is it that *they're*
the same?" (¶ 10.6). If, as were Adam and Blake, Fred thought of "they" as
"changing area" and "changing volume," then his confusion is understandable.
He was thinking of two things that are not the same. Second, if he was thinking
of changing area in relation to changing volume, then it is evident why he could
not visualize what is expressed in the formulation

$$\frac{V(h + \Delta h) - V(h)}{\Delta h} = \frac{A \times \Delta h}{\Delta h} \quad (\P\ 10.11) \ .$$

He was not thinking of the slab (Figure 20) as an accrual of volume – composed
multiplicatively of disk area and height – in comparison to a change in height.
Instead, Fred seemed to imagine the slab as that which defined the upper bound
of the water.

I sensed the confusion between "both changing" and one quantity having the
same value as the rate of change of the other, and attempted to refocus their attention on the ideas of rate of change of volume on the one hand and area of the disk
on the other hand. This exchange is given in Excerpt 11.

EXCERPT 11
11.1   Pat: Well, here, let's try this. What I hear is a little mixing of the ideas of area,
                     change in the area, and change in the volume. You're right that the volume
                     only gets bigger when the area gets bigger. But thinking of it that way ... I
                     don't see much hope in that giving us insight into why the rate of change of
                     the volume is actually the same as the area function. [*Pause.*] The idea that
                     as one gets bigger the other gets bigger doesn't seem to help much.

11.2   Blake: It doesn't mean that they necessarily have to be the same.

11.3   Fred: Is it something to do with this rate [*indicates vertical change in water level*]
                     being exactly the same as that rate [*indicates change in radius of circular cross
                     section; see Figure 19*]?

11.4   Pat: [*Pause.*] Uh ... I don't think they're the same.

11.5   Fred: The same ... proportion.

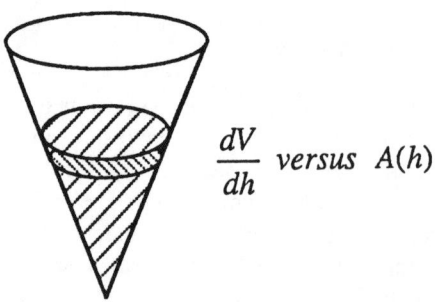

Fig. 21.   Figure 19, revised during discussion.

11.6   Blake: Yeah ... those segments are proportional, but the rates of change are different. [*Pause. Erases Figure 20.*] Here's what we're comparing. We're comparing [*writes "dV/dh"*] the rate of volume with respect to *h*, versus area as a function of *h* [*writes "vs. A(h)"*]. See, over here [*points to A(h)*] we're not talking about any kind of rate of change; we're just talking about area of a cross section as a function of its height from the bottom of the cone. Over here [*points to "dV/dh"*] we're talking about a rate of change. [*Long pause. Figure 20 is now erased from the board. Figure 19 is changed, appearing now as in Figure 21.*]

In Excerpt 11 I attempted to point out that what was the same (the graphs of cross-sectional area as a function of height and rate of change of volume with respect to height) were expressions of two different concepts – cross-sectional area and rate of change of volume. The ensuing discussion makes it evident that my formal expression of rate of change of one thing versus an amount of something else was not assimilated in the way I had intended.

EXCERPT 12

12.1   Alf: Am I thinking of this right. This [*points to "A(h)"*] is the area of the disk at some particular point [*moves hand up and down as if along vertical axis through middle of the cone*] ...

12.2   Pat: Yes, this is the area of a circular cross section.

12.3   Alf: At *h* [*points to "A(h)"*].

12.4   Pat: At ... [*moves hand vertically upward and then stops as if to show movement to a spot*] at *h*.

12.5   Alf: At *h* ... so ... then ... if you were thinking about this [*holds thumb and forefinger apart and next to top circular cross section in Figure 21, as if to measure its thickness*] ... the change in volume [*moves thumb and forefinger together, as if squeezing cylindrical slab, diminishing its height*] ... as delta *h* gets small, the change in volume ... delta *h* ... let me think.

12.6   Pat: Go ahead and express the change in the volume, and then the rate of change in the volume with respect to height.

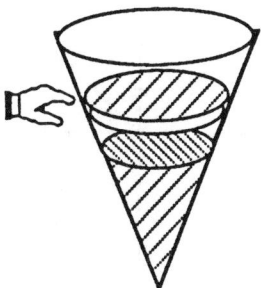

Fig. 22. Alf's addition to Figure 18. The hand depicted here represents Alf's hand; it was not part of Alf's drawing.

12.7	Alf:	Okay, the change in volume [*holds thumb and forefinger apart; long pause before he approaches diagram in Figure 21*] ... [*places thumb and forefinger slightly apart next to top circular cross section in Figure 21*] ... the change in the volume would be some minute [*minuscule*] ... distance in height ...
12.8	Pat:	Go ahead and draw it in.
12.9	Alf:	[*Draws new circular cross section; diagram now appears as in Figure 22.*] See ... I can almost picture in my mind that as delta h goes to zero [*moves thumb and forefinger together next to top circular cross section in Figure 22*] that that becomes the exact area disk that we're talking about. I mean that's ... that's ... In other words, as I shrink that height, this [*top of cylinder*] and this [*bottom of cylinder*] becomes [*pushes hands together one atop the other, as if squeezing something between them*] ... exactly that [*holds out one hand flat, parallel to floor, moving it side to side as if stroking the top of a table*] ... like a disk with no thickness. And when you write it out as a volume ... let's see, I don't remember the notation we used ... how did we write it ... it would be

$$\frac{V(h + \Delta h) - V(h)}{\Delta h}$$

12.10	Fred:	Equals.
12.11	Alf:	And this would be the change in volume.
12.12	Blake:	And you're saying that as delta h approaches zero, then we have, basically, ...
12.13	Alf:	I see that as *being* the area

Despite my attempt in Excerpt 11 and in Excerpt 12, ¶ 12.6, to orient students to think about rate of change of volume, Alf persisted in thinking about an increment in volume unrelated to any increment in height. Moreover, he began to think of a limiting process whereby, figurally, when you diminish the accrual's incremental thickness, you *get* an area. Alf seemed to be thinking of making the cylinder shorter and shorter, until top meets bottom. His image could be described formally as

$$\lim_{\Delta h \to 0} V(h + \Delta h) - V(h) = A(h),$$

which would have meant that

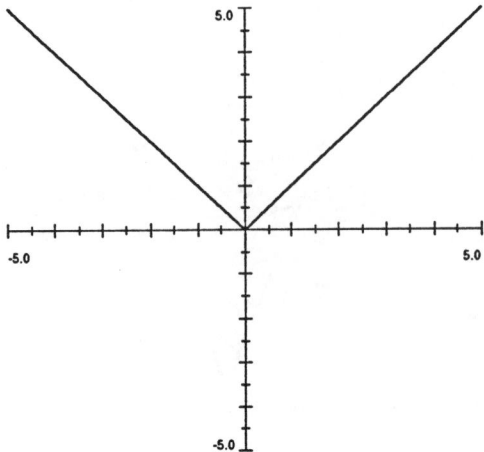

Fig. 23.   Graph to accompany follow-up assessment item 2.

$$\lim_{\Delta h \to 0} \frac{V(h + \Delta h) - V(h)}{\Delta h} = \lim_{\Delta h \to 0} \frac{A(h)}{\Delta h},$$

an equality I cannot interpret. The "Δh" in the denominator of Alf's difference quotient seemed insignificant to him. Perhaps this was because his focus was on the accrued "chunk" instead of on a meaning for the difference quotient that defined the function whose graph raised the issue in the first place.

There is a very natural interpretation of $[V(h + \Delta h) - V(h)]/\Delta h$ in regard to rate of change of accumulation. It is that it represents the average rate of change of volume over the interval $[h, h + \Delta h]$, where volume is *defined* by the value of the Riemann sum. Over the interval $[h, h + \Delta h]$, volume accrues by "stretching vertically" the cylinder having base area $A(h)$ – the area of the cross-sectional disk at height h.[13] Since the base area of the cylinder is constant over $[h, h + \Delta h]$, the volume grows at the rate $A(h)$. This is analogous to the case of speed. If we are considering a total accumulation of distance as a function of time, and if we assume that over some increment of time the distance is accruing at a constant rate, then regardless of how distance has accumulated prior to this increment of time, the *total* accumulated distance is changing at that constant rate over this increment of time.

Follow-up Assessment

Students took an exam two meetings after the end of the teaching experiment. I included four items to clarify possible sources of difficulty – two items on interpreting a difference quotient and two items on Riemann sums as functions. The difference quotient items were:

TABLE IV

Students' responses to Test Item 3. Note: Responses regarding information and responses regarding unit do not necessarily correspond within rows. The columns are presented independently of one another

Information given by $x(t)$		Unit	
Response	Frequency	Response	Frequency
Ave. rate of change of volume	4	Cubic meters per hour	7
Derivative	6	Hours	2
Rate of change of cooling	5	Degrees/hour	3
Average volume	1	0.1	1
Average change in volume	1	Square meters	1
Surface area	1	Volume/time/time	1
No answer	1	Other	4

2. a.   The graph in Figure 23 is of $f(x) = |x|$, $-5 \leq x \leq 5$. Sketch a graph of $h(x) = [f(x + \Delta x) - f(x)]/\Delta x$ over the same domain with $\Delta x = 0.5$. Use the coordinate system provided in the graph. *Hint: Imagine a sliding interval.*

   b.   Suppose you let Δx become progressively smaller. Explain what happens to the graph of $h(x)$.

3. a.   The volume in cubic meters of a cooling object t hours after removing a heat source is given by the function $v(t)$. Suppose a function $x(t)$ is defined as

   $$x(t) = \frac{\nu(t + 0.1) - \nu(t)}{0.1} \ .$$

   State precisely what information $x(t)$ gives about this object. (That is, don't tell me what $x(t)$ approximates. Tell we what information it actually gives.)

   b.   What is the unit of $x(t)$?

On Test Item 2 (difference quotient of absolute value function) 17 of 19 students drew a graph of the derivative of $|x|$. Only two students attended to the behavior of the function between -0.5 and 0. In follow-up interviews of each student, the 17 who drew a graph of the derivative of $|x|$ admitted thinking "derivative". The two who attended to $h(x)$'s behavior around 0 did not think of a rate of change or slope of a secant, but instead evaluated the function at different values of x and just happened to try values between -0.5 and 0.

The results of Item 3 are given in Table IV. Four students referred to an average rate of change of volume. Two others referred to an average, but not an average rate. Six students said that $x(t)$ is a derivative; five said that it is a rate of change, but of cooling. Only seven students gave an appropriate unit for $x(t)$.

The Riemann sum items were:

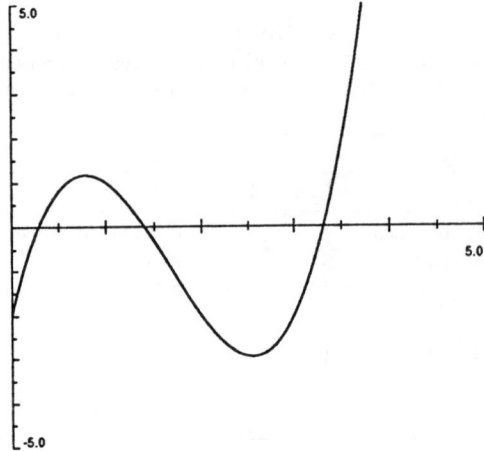

Fig. 24.   Graph to accompany follow-up assessment item 4.

4.   a.   The graph in Figure 24 is of a function $q(x)$ defined over the interval $[0, 5]$. Sketch a
          graph of

$$z(x) = \sum_{i=1}^{n} q\left(i\frac{x}{n}\right)\frac{x}{n} \quad \text{for } n = 1000 \text{ and } x \text{ in } [0, 5].$$

Use the coordinate system provided in the graph.

   b.   For what values of x (approximately) will $z(x)$ achieve a local maximum or a local
        minimum? Explain.

6.   Let $q(t)$ be defined by

$$q(t) = \sum_{i=1}^{t/\Delta t} f(i\Delta t)\Delta t.$$

Explain the process by which the expression

$$\sum_{i=1}^{t/\Delta t} f(i\Delta t)\Delta t$$

assigns a value to $q(t)$ for each value of t in the domain of f.

Responses to Item 4 were difficult to interpret. Eight students sketched appropriate graphs. They claimed to have identified the Riemann sum as "area" and to
have proceeded from that basis. Interviews with each student revealed a variety
of reasons for inappropriate graphs. One student said "derivative" just popped
into his head; several said that they didn't know how to proceed when they didn't
know what the actual function was (i.e., they did not have an analytic definition
of the function). Another student thought he should try to sketch a graph of the
areas of each of the 1000 rectangles you would get for $z(5)$.
Responses to Item 6 showed that the coordination of images involved in

understanding Riemann sums as functions was a complex act. One student wrote:

First the value of a certain chunk is measured by $i\Delta t$. This is then multiplied by the change which is Δt. This is repeated for every value of t and then added up. Each value of t is cut up into $t/\Delta t$ intervals, and added. $t/\Delta t$ is the number of intervals the piece is to be divided up into.

This student evidently had a number of problems, one being that he was imagining a "chunk" of a quantity independently of the analytic expression that established its measure — $i\Delta t$ does not "measure" the chunk, it just puts you at the right place to make it. The expression $f(i\Delta t)\Delta t$ gives the chunk's measure. A more serious problem, though, is that this student appeared to be imagining t and i varying simultaneously instead of as first i varying from 1 to $t/\Delta t$ for a fixed value of t and then varying t.

Another student wrote:

– Here Δt represents the size of each interval that f is being broken up into.
– So $t/\Delta t$ equals the number of intervals the graph of f is broken up into.
– So our i starts out at 1 and then goes to $t/\Delta t$.
– The expression first finds f and then it finds the ith interval of f that we are dealing with. Then it finds the value of the function f at that interval and then multiplies by Δt. This finds the area of that particular rectangle. Then we add it to the previous areas found and plot that point. You then connect all the points to get your curve.

The first sentence in this student's explanation, "...the size of each interval that f is being broken up into," suggests that she was imagining a Riemann sum over a fixed interval, which would normally correspond to an approximation of a definite integral $\int_a^b f(t)dt$ instead of the indefinite integral $\int_a^x f(t)dt$. Her last three sentences suggest that she, too, sometimes imagined i and t varying simultaneously.

Seven of 19 students expressed an appropriate order of variation for the index variable of the Riemann sum and the argument of the function. Five students appeared to have mixed images of definite and indefinite integrals. The remaining seven students had confounded the two variations so that everything was happening at once.

DISCUSSION

I structured the teaching experiment so that students were presented with a phenomenon requiring explanation: That when they graphed a function $f(x)$, defined the Riemann sum $g(x)$ as

$$g(x) = \sum_{i=1}^{n} f\left(i\frac{x}{n}\right)\frac{x}{n}$$

then graphed the function

$$Dg(x) = \frac{g(x + \Delta x) - g(x)}{\Delta x},$$

the graphs of $g(x)$ and $Dg(x)$ appeared identical for suitably large n and suitably small Δx. The functions were grounded in concrete settings, and explanations attempted by students' drew from their images and conceptions of the settings. My discussion will have three parts: Students' images as expressed during the teaching experiment and their contribution to students' difficulties, issues of notation, and implications of the present teaching experiment for standard approaches to the Fundamental Theorem and introductory calculus in general.

Students' Images

There seemed to be a confluence of images behind students' difficulties in construction and explanation for the problem of explaining an apparent relationship between $f(x)$, $g(x)$, and $Dg(x)$ as defined in the previous paragraph. These have to do with their images of function, their fixation on accrual as a solitary object, and a weak scheme for average rate of change. I conclude this section by relating the teaching experiment to Piaget's levels of imagery.

Images of Function

Students repeatedly made remarks that suggested a figural image of function – an image of a short expression on the left and a long expression on the right, separated by an equal sign (Thompson, in press b). This was not the only image students could conjure, but it seemed to be many students' "working image" – what came to mind without conscious effort whenever "function" was mentioned. This often oriented them away from grappling with conceptual connections entailed in situations dealing with covarying quantities.

In reviewing my notes and students work on assignments in Phase I, I noticed that students' explanations of the behavior of functions often spoke of the function's behavior as if it could be analyzed independently of its argument. Remarks were oriented to "the function" (often meaning the visual object called its graph) and not to a covariation of two variables. The analyses often referred to just one thing varying, this thing called "the function." Difficulties caused by an orientation to function as an idea with no interior showed up especially clearly when the function to be analyzed was a composition of functions. In analyzing the behavior of $f(g(x))$ it is critical to take into account the behavior of $g(x)$ in relation to x, for the variation of $g(x)$ is the variation of f's argument.

Finally, it seems that students' images of Riemann sums were insufficient to support their reasoning about a sum's rate of change. I suspect they were thinking of a Riemann sum as being static – that even though its argument could change, and the Riemann sum could be evaluated with a new argument, it was still a sum of unvarying "chunks" and a change in its argument was more like substituting a new value for the argument than a continuous change in its value. Their images of a Riemann sum seem not to have entailed a sense of motion, either in its argument or in its value.

Accruals as Solitary Objects

Students' remarks regarding a relationship between the *rate of change of a Riemann sum* and a *constituent quantity* in an accrual to the sum[14] always focused on the accrual as a solitary object (see especially Excerpt 12). To see a relationship between the two they needed either to conceptualize the accrual as itself accruing at a constant rate with respect to the independent quantity (e.g., height) or to conceptualize it as the average rate of increase in the accumulation over an increment in the independent quantity. In either case, it is necessary to have clearly in mind that accruals to the sum are constructed multiplicatively. In the first case, the accrual itself accruing at a constant rate, the accumulative quantity must be imagined to be constructed incrementally, where each increment is made by an increase in the quantity at a constant rate of change. In the second case, the accruals coming in "chunks", each accrual must be imagined to be a multiplicative combination of quantities (e.g., area and length) that will have increased at an *average* rate of change.

Students' fixation on accrual as a solitary object – simply as a thing with no constituent quantities – resembles young children's difficulties in constructing speed as a rate of change of distance with respect to time. Young children tend to think of speed as a distance – a measuring stick by which to measure other distances (Thompson and Thompson, in press; Thompson, in press a; Thompson and Thompson, 1992), and not something that grows in relation to a growing duration. This is not to say that the students in this teaching experiment understood speed in the same way as young children. Rather, it suggests that their schemes for rate and average rate were not operational to the extent that they could assimilate any covariate change to them.

Scheme for Average Rate of Change

A final source of difficulty, to which I already alluded in the previous section, was that students apparently did not have operational schemes for average rate of change. What do we mean by average rate of change of a quantity? We typically mean that if a quantity were to grow in measure at a constant rate of change with respect to a uniformly changing quantity, then we would end up with the same amount of change in the dependent quantity as actually occurred. An average speed of 55 km/hr on a trip means that if we were to repeat the trip traveling at a constant rate of 55 km/hr, then we would travel precisely the same amount of distance in precisely the same amount of time as had been the case originally. This notion is highly related to the Mean Value Theorem for derivatives, which says, in effect, that all differentiable functions do have an average rate of change over an interval and it is equal to some instantaneous rate of change within that interval. In the case of a Riemann sum, the rate of change of the sum for x within an interval $[q, q + \Delta q]$ is equal to the average rate of change of the quantity $f(t)\Delta t$ for some t in $[q, q + \Delta q]$ and for Δt varying from q to $q + \Delta q$ – which is just $f(t)$.

Coordination of Actions

As noted in the introduction, Piaget characterized his second level of imagery as, "In place of merely representing the object itself, independently of its transformations, this image expresses a phase or an outcome of the action performed on the object. ... [but] the image cannot keep pace with the actions because, unlike operations, such actions are not coordinates one with the other" (Piaget, 1967, p. 295). This seems to capture the nature of some students' understanding of Riemann sum, and other students' understanding of Riemann sum in relation to rate of change. Some students had not come to coordinate the variations of upper limit of summation and the variations in the index of the summation; some students had not coordinated the actions of forming a sum and multiplicatively constructing an accrual to a sum. Other students had mastered both of these coordinations but could not coordinate that ensemble of actions with the action of comparing multiplicatively the growth in an accrual with growth in one of its constituent quantities. As Piaget said, their actions outpaced their images because their actions were not coordinated. Operational understanding of the Fundamental Theorem entails the coordination of these actions so that the scheme remains in balance. Operational understanding of the Fundamental Theorem allows one to hold simultaneously in relation to one another the mental actions of forming accruals, accumulating accruals, and comparing an accrual to one of its constituent quantities multiplicatively.

Notation

I should point out that the above discussion is colored by one serious matter. This is that students often acted from an orientation which led them to use notation opaquely. We discussed this tendency during class on several occasions. A common remark was that this seemed, from their point of view, the most efficient way to cope with what they thought had been expected of them, both in high school and in college. When students did interpret notation, it often came as an afterthought, and they often tended to read into the notation what they wanted it to say, without questioning how what they actually wrote might be interpreted by another person. More often, though, students would not interpret the notation with which they worked, but would instead associate patterns of action with various notational configurations and then respond according to internalized patterns of action. Their orientation toward notational opacity, while having nothing to do with conceptual difficulties with the Fundamental Theorem of Calculus as such, certainly contributed to their not having grappled with key connections.

Implications for Contemporary Treatments of the Fundamental Theorem

The approach taken within this teaching experiment resembles Anton's (1992, pp. 320–323) intuitive development of the Fundamental Theorem, with the exception that Anton does not employ Riemann sums and focuses exclusively on the case of area bounded by a function's graph. Anton's intuitive development is not oriented at students' conceptualizing the Fundamental Theorem so much as to motivate his upcoming focus on techniques of antidifferentiation.[15] A focus on techniques of antidifferentiation is historically accurate – Newton's and Leibniz' motivation for constructing the Fundamental Theorem was so that they could make algorithmic the process of constructing analytic expressions for areas under curves. However, Anton switches, unannounced, to another conceptualization in justifying the Fundamental Theorem – he bases it on the mean value theorem for integrals. The mean value theorem for integrals says that continuous functions have an average value over an interval, where the average value f_{av} over $[a, b]$ of a continuous function f is defined as $f_{av} = (b - a)^{-1} \int_a^b f(x)dx$ (Swokowski, 1991, p. 281). On the other hand, the mean value theorem for derivatives says that if f is continuous on a closed interval $[a, b]$ and differentiable on the open interval (a, b), then there exists a number c in (a, b) such that $f'(c) = [f(b) - f(a)]/(b - a)$ (Swokowski, 1991, p. 179). The mean value theorem for integrals allows a formal proof of the Fundamental Theorem to go smoothly – we can substitute the average value of the integrated function for the integral of the function over the increment in its argument. On the other hand, the mean value theorem for derivatives supports a conceptualization of what is going on – the accumulation (integral) of the multiplicatively-constructed quantity $f(t)dt$ is changing at an average rate of change that is equal to $f(t)$ for some t in $[x, x + \Delta x]$.

A typical proof of the Fundamental Theorem goes something like this: Let $f(x)$ be a continuous function defined on $[a, b]$. Define $F(x)$ as $F(x) = \int_b^x f(t)dt$. Then

$$F'(x) = \lim_{h \to 0} \frac{\int_a^{x+h} f(t)dt - \int_a^x f(t)dt}{h}$$

$$= \lim_{h \to 0} \frac{\int_x^{x+h} f(t)dt}{h}$$

$$= \lim_{h \to 0} \frac{f(z)h}{h} \text{ for some } z \in [x, x + h].$$

The last line is where the mean value theorem for integrals is used. The integral $\int_x^{x+h} f(t)dt$ is equal to $f(z)h$ for some z in the interval $[x, x + h]$. That is, the integral is equal to the average value of the function over the interval times the width of the interval. Then, as $h \to 0$, $z \to x$, and so $F'(x) = f(x)$.

The problem with the typical proof is not so much in the proof as that it is presented as modeling a static situation. It is presented in such a way that nothing is *changing*. If students are to understand $F'(x)$ is a rate of change, then something must be changing. But as soon as we bring in the idea of motion, then the mean value theorem of integrals becomes a conceptual misfit – it doesn't fit the image of $\int_a^x f(t)dt$ as a dynamic accumulation of a quantity. We must rely on the mean value theorem for derivatives to support the idea of rate of accumulation.

However, this teaching experiment suggests that a great deal of image-building regarding accumulation, rate of change, and rate of accumulation must precede their coordination and synthesis into the Fundamental Theorem.

NOTES

[†] Research reported in this paper was supported by National Science Foundation Grants No. MDR 89-50311 and 90-96275, and by a grant of equipment from Apple Computer, Inc., Office of External Research. Any conclusions or recommendations stated here are those of the author and do not necessarily reflect official positions of NSF or Apple Computer. Also, I wish to thank Paul Cobb and Guershon Harel for their helpful reactions to an earlier draft of this article.

[1] Tom Kieren and Susan Pirie (Kieren and Pirie, 1990, 1991; Kieren, 1989; Pirie and Kieren, 1991) make it evident that the act of imagining can itself inform our images.

[2] The Latin root of "confused" is *confundere*, to mix together. Thus, one way to think of being in a state of confusion is that we create inconsistent images while operating.

[3] Winograd and Flores (1986) give similar criticisms of referential meaning in cognitive science.

[4] I should point out that when students speak of "rate" as in "distance equals rate times time," they need not be speaking of anything having to do with rate as I use the term. They may be engaging in mere "symbol speak," having no imagistic content except for the imagery of notational actions (Hayes, 1973).

[5] This is a nonstandard interpretation. I am actually anticipating discussions regarding Newton's development of the Fundamental Theorem.

[6] Here I must stress that I am talking about images and not about logical demonstration. The notion of accrual, when made rigorous, poses many problems regarding continuity of change and relationships between discrete and continuous quantities (this is the well-known problem of infinitesimals). But that is beside the present point – what sorts of images make the Fundamental Theorem *intelligible*.

[7] We must keep in mind that during Newton's time all functions were thought to be continuous and differentiable almost everywhere. It was only later that pathological functions and Fourier series showed that these ideas could be pushed beyond a point where they became insufficient as a foundation for the calculus (Kuhn, 1970; Wilder, 1967, 1968).

[8] I must stress once more that this is not a rigorous development. Rather, it is about images that might support the "obviousness" of the Fundamental Theorem. Also, it seems that Newton sensed the inadequacies of infinitesimals as a logical foundation for his calculus and eventually disavowed them (Boyer, 1959, p. 213). Nevertheless, it seems clear that his initial insights were facilitated by his acceptance of infinitesimals.

[9] It is important to note that, formally, the unit of $\Delta T / \Delta V$ should be $hr/(mi/hr)$, but Sue evidently reasoned that $1/\Delta V ths$ of the total change in velocity should correspond to $1/\Delta V ths$ of the time in which the change in velocity occurred. Therefore each increment of the total time would be $\Delta T / \Delta V ths$ of one hour This is the kind of reasoning about rates depicted in Figures 1 and 2.

[10] In a later problem, "about how far does a rock fall on the moon in its fourth second of falling if on the moon falling things speed up at the rate of 6 ft/sec every second," Sue concluded that at the beginning of the fourth second the rock would be falling 18 ft/sec, and that each one-tenth of a second thereafter the rock would speed up by 0.6 ft/sec.

[11] A more accurate representation of the Riemann sum would be to have $[x/\Delta x]$, the greatest integer less than or equal to $x/\Delta x$, in the upper limit of the summation. However, our graphing program used the convention that the upper limit of a summation is truncated to an integer, so it was only necessary to put $x/\Delta x$ as the upper limit of the summation.

[12] My class presentations were with a Macintosh Powerbook connected to an LCD projection panel. I used Theorist, which allows expressions and functions to be displayed in standard mathematical notation and which allows graphs, diagrams, etc. to be placed anywhere on the computer screen. The function definitions and graphs presented here are taken directly from my class presentation.

[13] See the discussion of Figure 3.

[14] By an "accrual" to a Riemann sum I mean the thing whose measure is $f(t_i)\Delta t$. So by "constituent quantity" I mean the thing measured by $f(t_i)$ in $f(t_i)\Delta t$.

[15] An antiderivative of $f(x)$ is a function $g(x)$ such that $g'(x) = f(x)$. Antidifferentiation is the process of finding an antiderivative.

REFERENCES

Anton, H.: 1992, *Calculus with Analytic Geometry*, John Wiley and Sons, New York.
Baron, M.: 1969, *The Origins of Infinitesimal Calculus*, Pergamon Press, New York.
Boyd, B. A.: 1992, *The Relationship between Mathematics Subject Matter Knowledge and Instruction: A Case Study*, Masters Thesis, San Diego.
Boyer, C. B.: 1959, *The History of the Calculus and Its Conceptual Development*, Dover, New York.
Cobb, P. and von Glasersfeld, E.: 1983, 'Piaget's scheme and constructivism', *Genetic Epistemology* 13, 9–15.
Courant, R.: 1937, *Differential and Integral Calculus*, Interscience, New York.
Dewey, J.: 1929, *The Sources of a Science of Education*, Liveright Publishing, New York.
Dubinsky, E.: 1991, 'Reflective abstraction in advanced mathematical thinking', in D. Tall (ed.), *Advanced Mathematical Thinking*, pp. 95–123, Kluwer, Dordrecht, The Netherlands.
Goldenberg, E. P.: 1988, 'Mathematics, metaphors, and human factors: Mathematical, technical, and pedagogical challenges in the educational use of graphical representation of functions', *Journal of Mathematical Behavior* 7, 135–173.
Hayes, J. R.: 1973, 'On the function of visual imagery in elementary mathematics', in W. G. Chase (ed.), *Visual Information Processing*, pp. 177–214, Academic Press, New York.
Johnson, M.: 1987, *The Body in the Mind: The Bodily Basis of Meaning, Imagination, and Reason*, University of Chicago Press, Chicago, IL.
Kaput, J. J.: in press, 'Democratizing access to calculus: New routes to old roots', in A. H. Schoenfeld (ed.), *Mathematics and Cognitive Science*, Mathematical Association of America, Washington, D.C.
Kieren, T. and Pirie, S.: 1990, April, 'A recursive theory for mathematical understanding: Some elements and implications', Paper presented at the Annual Meeting of the American Educational Research Association, Boston, MA.
Kieren, T. and Pirie, S.: 1991, 'Recursion and the mathematical experience', in L. P. Steffe (ed.), *Epistemological Foundations of Mathematical Experience*, pp. 78–101, Springer-Verlag, New York.
Kieren, T. E.: 1989, 'Personal knowledge of rational numbers: Its intuitive and formal development', in J. Hiebert and M. Behr (eds.), *Number Concepts and Operations in the Middle Grades*, pp. 162–181, National Council of Teachers of Mathematics, Reston, VA.
Kosslyn, S. M.: 1980, *Image and Mind*, Harvard University Press, Cambridge, MA.
Kuhn, T. S.: 1970, 'Logic of discovery or psychology of research', in I. Lakatos and A. Musgrave (eds.), *Criticism and the Growth of Knowledge*, pp. 1–22, Cambridge University Press, Cambridge, U.K.
Piaget, J.: 1950, *The Psychology of Intelligence*, Routledge and Kegan-Paul, London.
Piaget, J.: 1967, *The Child's Concept of Space*, W. W. Norton, New York.
Piaget, J.: 1968, *Six Psychological Studies*, Vintage Books, New York.
Piaget, J.: 1971, *Genetic Epistemology*, W. W. Norton, New York.
Piaget, J.: 1976, *The Child and Reality*, Penguin Books, New York.
Piaget, J.: 1980, *Adaptation and Intelligence*, University of Chicago Press, Chicago.
Pirie, S. and Kieren, T.: 1991, April, 'A Dynamic Theory of Mathematical Understanding: Some features and implications', Paper presented at the Annual Meeting of the American Educational Research Association, Chicago, IL.
Sfard, A.: 1991, 'On the dual nature of mathematical conceptions: Reflections on processes and objects as different sides of the same coin', *Educational Studies in Mathematics* 22, 1–36.
Steffe, L. P.: 1991, 'Operations that generate quantity', *Journal of Learning and Individual Differences* 3, 61–82.
Steffe, L. P.: in press, 'Children's multplying and dividing schemes: An overview', in G. Harel and J. Confrey (eds.), *The Development of Multiplicative Reasoning in the Learning of Mathematics*, SUNY Press, Albany, NY.

Swokowski, E. W.: 1991, *Calculus*, PWS-Kent, Boston, MA.

Tall, D.: 1986, *Building and Testing a Cognitive Approach to the Calculus Using Interactive Computer Graphics*, Doctoral dissertation, University of Warwick.

Tall, D., Van Blokland, P., and Kok, D.: 1988, *A Graphic Approach to the Calculus*, Computer Program for IBM and Compatibles, Warwick University, Warwick, U.K.

Tall, D. and Vinner, S.: 1981, 'Concept images and concept definitions in mathematics with particular reference to limits and continuity', *Educational Studies in Mathematics* 12, 151–169.

Thompson, A. G. and Thompson, P. W.: in press, 'Talking about rates conceptually: A teacher's struggle', *Journal for Research in Mathematics Education*.

Thompson, P. W.: 1985, 'Experience, problem solving, and learning mathematics: Considerations in developing mathematics curricula', in E. Silver (ed.), *Teaching and Learning Mathematical Problem Solving: Multiple Research Perspectives*, pp. 189–243, Erlbaum, Hillsdale, NJ.

Thompson, P. W.: 1991, 'To experience is to conceptualize: Discussions of epistemology and experience', in L. P. Steffe (ed.), *Epistemological Foundations of Mathematical Experience*, pp. 260–281, Springer-Verlag, New York.

Thompson, P. W.: in press a, 'The development of the concept of speed and its relationship to concepts of rate', in G. Harel and J. Confrey (eds.), *The Development of Multiplicative Reasoning in the Learning of Mathematics*, SUNY Press, Albany, NY.

Thompson, P. W.: in press b, 'Students, functions, and the undergraduate mathematics curriculum', *Research in Collegiate Mathematics Education* 1.

Thompson, P. W. and Thompson, A. G.: 1992, April, 'Images of rate', Paper presented at the Annual Meeting of the American Educational Research Association, San Francisco, CA.

Vinner, S.: 1987, 'Continuous functions: Images and reasoning in college students', in *Proceedings of the Annual Meeting of the International Group for the Psychology of Mathematics Education*, PME Montréal, Canada.

Vinner, S.: 1989, 'Avoidance of visual considerations in calculus students', *Journal of Mathematical Behavior* 11, 149–156.

Vinner, S.: 1991, 'The role of definitions in the teaching and learning of mathematics', in D. Tall (ed.), *Advanced Mathematical Thinking*, pp. 65–81, Kluwer, Dordrecht, The Netherlands.

Vinner, S.: 1992, 'The function concept as a prototype for problems in mathematics learning', in G. Harel and E. Dubinsky (eds.), *The Concept of Function: Aspects of Epistemology and Pedagogy*, pp. 195–214, Mathematical Association of America, Washington, D.C.

Vinner, S. and Dreyfus, T.: 1989, 'Images and definitions for the concept of function', *Journal for Research in Mathematics Education* 20, 356–366.

von Glasersfeld, E.: 1978, 'Radical constructivism and Piaget's concept of knowledge', in F. B. Murray (ed.), *Impact of Piagetian Theory*, pp. 109–122, University Park Press, Baltimore.

Wilder, R.: 1967, 'The role of axiomatics in mathematics', *American Mathematical Monthly* 74, 115–127.

Wilder, R.: 1968, *Evolution of Mathematical Concepts: An Elementary Study*, Wiley, New York.

Wilder, R.: 1981, *Mathematics as a Cultural System*, Pergamon Press, New York.

Winograd, T. and Flores, F.: 1986, *Understanding Computers and Cognition: A New Foundation for Design*, Ablex, Norwood, NJ.

Patrick W. Thompson,
Center for Research in
Mathematics and Science Education
San Diego State University,
6475 Alvarado Rd. #206,
San Diego, CA 92120, USA

JÖRG VOIGT

NEGOTIATION OF MATHEMATICAL MEANING AND LEARNING

MATHEMATICS

ABSTRACT. The teaching-learning process is considered as a social interaction. In this microethno-graphical case study an elementary teacher and first graders are observed when they ascribe mathe-matical meanings of numbers and of numerical operations to empirical phenomena. Because of the differences of their ascriptions, the teacher and the students negotiate mathematical meanings. Also interactional regularities help the participants to cope with ambiguity. According to different theoret-ical approaches, the text discusses some indirect relations between social interaction and mathematics learning. Several classrooms episodes are interpreted to illustrate specific theoretical concepts.

1. INTRODUCTION

In several countries there is a growing body of research that supports the relevance of social activities to learning mathematics. In Brazil, Carraher et al. (1985) and Saxe (1990) study relations between cultural activities and cognitive develop-ment by comparing the children's mathematical thinking outside and inside the culture of school. In Great Britain, Bishop and Goffree (1986) and Walkerdine (1988) analyze the social constitution of mathematical meaning during classroom processes. In France, Balacheff and Laborde (1988) develop experimental situa-tions in which students interactively constitute mathematical solutions. In Italy, Bartolini-Bussi (1990) introduces phases of mathematical discussions into class-rooms and explores the learning processes taking place during the discussions. In the United States, Cobb et al. (1991) analyze how the change of social norms in the classroom affects the students' learning. In Germany, Bauersfeld et al. (1988) investigate the negotiation of mathematical meanings and the students' and the teacher's experience during these processes.

Although the list is not complete it gives an impression of the change of many mathematics educators' foci of attention. In the decades before, clinical interviews were conducted in order to understand the individual student's thinking. Presently, social dimensions are not excluded or neglected, but they are taken into account by undertaking "ethnographical" case studies in order to understand mathematics learning in usual contexts. The basic assumption is that social dimensions are not the peripheral conditions of learning mathematics but are intrinsic to learning mathematics.

Many of the present approaches are extensions of psychological lines of thought. The strong reference to psychology is surprising because the teaching-learning process implies an interpersonal relationship, and interpersonal relation-ships are conceptualized and studied typically in sociology, a discipline which is

Educational Studies in Mathematics **26**: 275–298, 1994.
© 1994 *Kluwer Academic Publishers. Printed in the Netherlands.*

rich of useful theories, methodological programs, etc. The following text proposes
an interactionist perspective on teaching and learning mathematics that refers to
a specific domain of sociology which is epistemologically compatible with the
constructivist psychological perspective.

On the one hand, the text claims to present a theoretical perspective. On the
other hand, the theoretical concepts presented were elaborated while classroom
scenes were analyzed in detail. Therefore, the text will combine abstract ideas
and concrete classroom episodes. Although every classroom situation is open
to various interpretations, in the next sections the possible interpretations are
restricted in order to illustrate the theoretical concepts.

The case study of a German elementary classroom forms the empirical background. The author has observed and videotaped every third lesson of an elementary class from the beginning of first grade to the middle of second grade.
The case study is conducted with regard to the standards of microethnography (cf.
Voigt, 1990).

One objective of the project is to reconstruct how the teacher and the students ascribe mathematical meanings to empirical phenomena (i.e., pictures, story
problems, etc.) presented in the classroom. The focus of attention is on the
processes of mathematization, that is, on what, for the observer, is the transformation of empirical situations into mathematical statements, and vice versa.
Referring to Steinbring's epistemological analysis (1991) mathematical meaning
can be viewed as related to empirical facts as well as to the mathematical symbols
(signs). Especially in first grade, school mathematical knowledge is inseparable
from the students' empirical experiences.

2.. ASCRIBING MATHEMATICAL MEANINGS TO EMPIRICAL PHENOMENA

Mathematical Meaning from a Sociological Point of View

Research in mathematics education takes a keen interest in understanding the
development of mathematical meaning. But what is the meaning of "mathematical meaning"? Answers depend on various researchers' perspectives. Several
lines of research in cognitive psychology emphasize the individual's sense making processes and the subject's cognitive development. Several philosophical
perspectives emphasize the development of mathematical knowledge and its epistemological status, assuming that the nature of mathematics implies meanings
independently of the inner world of persons. In the following discussion, the
development of mathematical meaning is studied from a sociological perspective.
Mathematical meaning is taken as a product of social processes, in particular as
a product of social interactions. From this point of view, mathematical meanings
are primarily studied as emerging between individuals, not as constructed inside
or as existing independently of individuals.

The sociological perspective does not fit the widely held opinion that a single
isolated mathematician could discover mathematics as an eternal knowledge or
could recollect it as an inborn knowledge (see platonism, discussed in Struve and

Fig. 1.

Voigt (1988)). However, philosophers like Lakatos (1976) or Wittgenstein (1967) emphasize the argumentation processes and "language games" among persons when they explain the development of mathematical meanings. These philosophical works support the sociological assumption that mathematical meaning can be studied as emerging in social relationships among individual subjects.

The text presents a specific sociological approach based on microsociology, with particular emphasis on symbolic interactionism (Blumer, 1969; Goffman, 1974) and on ethnomethodology (Garfinkel, 1967; Mehan, 1979). However, the sociological concepts have been modified in order to deal with teaching and learning mathematics (Bauersfeld et al;, 1988).

Ambiguity

In order to view mathematical meanings as a matter of negotiation, it is helpful to take into account the ambiguity of objects in the mathematics classroom. According to folk beliefs, in mathematics lessons, the tasks, the questions, the symbols (signs), etc., have definite, clear-cut meanings. If one looks at microprocesses in the classroom carefully, those matters seem to be ambiguous and call for interpretation. What is the meaning of the numeral 5 for young children in a specific situation? Is the meaning bound to concrete things (e.g., the little finger of my left hand), does the symbol remind the child of previous activities (e.g., a difficult number to write), does it activate specific emotions (e.g., my favourite number), does it derive its meaning as related to other numbers (e.g., equal to 2 plus 3, 1 and 4, 0 and 5), etc.?

The following is an example of various interpretations of a picture which is presented as an assumed unambiguous task in a regular German schoolbook for first graders (see Figure 1).

Several children and adults were asked to give the correct number sentence (Voigt, 1990). Solutions to the task are given below:

– $2 + 3 = 5$ (sum of the bananas)
– $5 - 2 = 3$ (the keeper gives two bananas to the ape)
– $1 + 1 = 2$ (the keeper and the ape)
– $3 - 2 = 1$ (the keeper has one banana more than the ape)
– $5 - 4 = 1$ (one banana more than hands, the keeper will lose the middle one)

One of the findings of a research project (Neth and Voigt, 1991) was that in principle such pictures, text problems, games, and stories have multiple meanings if the children interpreting the task are not familiar with the specific type of the task. Nevertheless, many authors of textbooks and mathematics teachers seem to assume that these objects have unambiguous meanings and that the tasks have definite solutions. The processes of mathematization taken for granted become problematic when the situations are interpreted by subjects who are (still) not members of the classroom culture.

One assumption made by interactionists is that every object or event in human interaction is plurisemantic ("indexical", Leiter, 1980, pp. 106–138). In order to make sense of that object or event, the subject uses his/her background knowledge and forms a sensible context for interpreting the object. For example, a first grader can interpret the picture as the invitation to share a personal experience at a zoo; or the teacher can take the picture as the opportunity for motivating students to apply subtraction. The subject does not necessarily experience the ambiguous object as plurisemantic. Rather, if the background understanding is taken for granted, the subject experiences that object as factual.

Different Individual Perspectives

An ambiguity is an essential characteristic of the teaching-learning situation if it can not be eliminated in a single situation by agreement. Referring to Goffman's (1974) frame analysis, Krummheuer (1983) reconstructs different background understandings (framings) between the teacher and the students of an algebra class over an extended period of time. Comparing the interactions during collaborative learning with the interactions during frontal class teaching, he demonstrates that "misunderstandings" between the teacher and the students are quite typical. Steinbring's epistemological study (1991) explains why the disparity between the teacher's and the student's background knowledge exists necessarily when the teacher is introducing new mathematical concepts which require a change in the student's background understanding of the nature of mathematical concepts.

Especially at the elementary level, there is a strong tendency to minimize the ambiguity of mathematical symbols (signs) and of empirical issues which should serve as representations of mathematics. There is the hope that students can learn mathematics readily if mathematical meanings are definitely bound to specific dealings with concrete things. For example, in German schools subtraction is introduced as decreasing, say, the number of chips or apples by taking some away; initially, subtraction is not established as determining the difference between the number of chips taken away and the number of the remaining chips. The point

is not that the latter alternative should be preferred by the teacher in the introduction of subtraction. Rather, a student might prefer it in a specific situation.
So, the student's original interpretation of the situation can cause conflicts and
re-establish the ambiguity of the relationship of mathematical symbols (signs) to
concrete materials. The following episode provides an example.

The teacher has introduced the multiplication of numbers by repeated addition and by identifying
the first factor of the product as the multiplier and the second as the multiplicand, respectively. This
introduction is consistent with the textbook and represents common practice in German schools today.
In the course of this introduction, the teacher holds up three packs each containing ten pens. From the
teacher's perspective, these materials represent "3 times 10" – analogous to similar problems posed
before.

| Teacher: | To these packs, please, find a multiplication task. |
| Natalie: | 10 times 3. |

Student become noisy, teacher calls them to order.

Teacher:	Yeah, but, is this really correct?
Student A:	No!
Teacher:	Do you all agree?
Student B:	No.
Student C:	That's the same like a swap task.
	(known by the students in the context of addition, e.g., $3 + 4 = 4 + 3$)

An interpretation: At first Natalie does not fulfill the teacher's expectation
of the mathematization, that is, the first factor represents the amount of packs.
Possibly, Natalie's mental representation conflicts with the official mathematization previously constituted in the lesson. The teacher seems to elicit a negative
evaluation of Natalie's answer. Considering the "weaker" students, perhaps, the
teacher wants to avoid the risk of ambiguity. However, student C hints to the
mathematical identity of the different forms.

In this episode, presumably, the teacher and two students assume that each
number of a product is unambiguously represented by specific materials (packs
and pens, respectively). But Natalie offers a divergent meaning of a product,
and a student justifies its correctness by drawing an analogy to a mathematical
argument established previously in the context of addition. (Also note that other
multiplication tasks are imaginable, e.g., 3 times 2 deutschmarks if one pack costs
2 deutschmarks.)

Mixing Empirical and Theoretical Aspects

An epistemological consideration provides an additional explanation for why ambiguity is an essential feature of mathematics in school. In their epistemological
studies, Steinbring (1993) and Struve (1990) elaborate the nature of mathematics which is peculiar to school. On the one hand, a pure mathematician who
tries to discover truths about a mathematical reality can understand meanings

of mathematical symbols as unambiguous because of the formal interrelatedness among the objects in the context of a mathematical theory. Concerned about the meaning of a mathematical symbol, the mathematician would consider its formal definition and its theoretical use. This mathematician's experience of certainty does not depend on an empirical (physical) existence of the objects (see, for example, non-Euclidean geometries). On the other hand, especially in elementary school, the meanings of mathematical symbols (signs) are related to empirical issues (numerals to materials, geometrical terms to the physical space, etc.).

This problem corresponds to a distinction in the application of mathematics. Inferences within a mathematical model and the selected assumptions for modelling empirical phenomena have different rational bases. Even if the student was be able to draw the inferences, the student could not mathematically discover those assumptions intended by the teacher. The situation is even more complex because the student is not encouraged to experience this distinction if, as commonly practiced, the participants tend to identify empirical statements with theoretical ones – mixing different rational bases. If the teacher takes empirical phenomena as starting-points in order to make the students familiar with specific mathematical concepts and if a student mathematizes the empirical phenomena differently than expected by the teacher, a conflict is possible that can not be solved by pure inferences. This is one reason why mathematical meanings in school are necessarily a matter under negotiation.

Also if the participants agree on a specific mathematization, the negotiation of meaning can be helpful that distinguishes between empirical and theoretical reasons.

> Why does 2 plus 3 equal 5? Is it true because 5 bananas can be counted in the picture? Or, is it true because 2 plus 3 has to be one more than 2 plus 2 which equals 4 – independently of the bananas?

Although empirical plausibility can represent inner-mathematical reasons, the students should become familiar with the mathematical rationality in the long run. Therefore, the teacher must ensure that the students do not restrict their thinking to empirical evidence which is obvious to the students. Through processes of negotiation of what counts as a reason, the teacher can stimulate the students to develop a sense of theoretical reasoning even if empirical reasons are convincing and seem to be sufficient.

3. NEGOTIATION OF MATHEMATICAL MEANING

Implicit and Explicit Negotiation

From the view of symbolic interactionism, interaction is more than a sequence of actions and reactions. The participant of the interaction monitors his action in accordance with what he assumes to be the other participants' background understandings, expectations, etc. At the same time, the other participants make sense of this action by adopting what they believe to be the actor's background understandings, intentions, etc. The subsequent actions of the other participants are interpreted by the former actor with regard to his expectations and can prompt a reconsideration, and so on. For example, the student can interpret the teacher's reaction as a specific evaluation of his own thinking even though the teacher's reaction might only have been the expression of amusement at a particular moment during the student's action. Using his background knowledge of the teacher's supposed emotion coupled with his emotion of being ashamed, for example, the student might search for more advanced ways to interpret and to solve the problem at hand. Experiencing a similar situation, another student might cope with the situation by pleasing the teacher superficially. Of course, the teacher could react differently according to his background knowledge of the students' different dispositions. So, from the interactionist perspective, mathematical meanings are negotiated even if the participants do not explicitly argue from different points of view. Nevertheless, one can study the negotiation directly if we focus our attention on conflicts among the teacher and the students. In cases of conflict, the accomplishment of intersubjective meanings taken as mathematical meanings becomes problematic.

In the previous episode, the student Natalie presented a mathematical term which we can interpret as an indication that Natalie ascribed another meaning to a multiplication term not expected by the other participants. In any case, Natalie's mathematical term did not fit with the teacher's and other students' expectations. Then, Student C justified Natalie's statement using an argument which from the observer's theoretical point of view implies the commutative law. Thus, the teacher is under the obligation to consider this argument. The situation is tricky because the attempt to make the commutative law plausible on an empirical level would demand the establishment of a close relationship between the order of mathematical symbols in the term and the concrete materials. The teacher reacts in the following way:

Teacher: That's a swap task. If I empty the packs and if I line them up in threes, then this is 30, too. In lines of three! But to these packs (points to the three original packs) you would really have to write 3 times 10 ("müßtet ihr eigentlich schreiben"). And then you have the same amount of all pens.

The teacher mentions that the difference between the orders of pens does not affect the total amount. Implicitly, the teacher points out that the different terms have something in common at the level of the concrete things. The obligation that the students undertake only a specific mathematization (i.e., the amount of packs represents the first factor) seems to be weakened. Also, the obligation does not seem to be definite because in her instruction the teacher uses the subjunctive mood "irrealis." That is, the German phrase "müßtet ihr eigentlich schreiben" is sometimes used to express an expectation on which one does not insist any longer. The students can take the phrase as a hint that Natalie's interpretation is not wrong but not expected.

Mathematical Meanings Taken-to-be-Shared

Understanding the negotiation of meaning as the accomplishment of intersubjective meanings does not imply that the teacher and the students "share knowledge." From the symbolic interactionist and the radical constructivist points of view, only mathematical meanings taken-to-be-shared can be produced through negotiation. Goffman (cited in Krummheuer, 1983) and von Glasersfeld (cited in Cobb, 1990) use the term "working interim" and "consensual domain," respectively, in order to describe that the participants interact as if they interpret the mathematical topic of their discourse in the same way although no one can be safeguarded against a missing fit of subjective background understandings. One can never be sure that two persons are thinking similarly if they collaborate without conflict, especially if they agree about formal statements and processes. It is characteristic of formal mathematics that people can coordinate their actions smoothly while they are ascribing different meanings to the objects.

The observed teacher and students solve many formal mathematical problems after the introduction of mathematical concepts, such as multiplication. In these routine situations, the students can interpret the mathematical terms differently from the teacher's intention without finding wrong solutions. For example, Natalie could solve the formal multiplication problems correctly even if she is not able to accept the teacher's opposing interpretations of the factors and even if she does not understand the equality as representing the commutative law.

In the classroom, the participants constitute mathematical meanings as taken-to-be-shared interactively. What is meant by taken-to-be-shared emerges during processes of negotiation. From the observer's point of view, taken-to-be-shared is not a (partial) match of individuals' constructions; it is not a cognitive element but has its own right at the level of interaction. "Symbolic interactionism views meaning ... as arising in the process of interaction between people. The meaning of a thing grows out of the ways in which other persons act toward the person with regard to the thing. ... – symbolic interactionism sees meanings as social products" (Blumer, 1969, p. 5).

In the episode described above, the meaning of a swap task also emerges in the context of multiplication. The participants speak of a swap task as if they ascribe the same meaning to it. Presumably, the teacher did not intend to mention the commutative law. (According to the text book it should become a topic several lessons later.) Also, it is not clear whether the other students can understand the argument which has been presented by Student C. However, the subsequent interactions confirm the assumption that the participants arrive at a taken-to-be-shared understanding of the swap task (the commutative law of multiplication); they act as if they use it implicitly without provoking a conflict. Later in the same lesson, even the teacher transcends her previous rules when she interprets "15 times 2" as 2 packs, each containing 15 pens.

Fig. 2.

Mathematical Theme

In the course of negotiation, the teacher and the students (or the students among themselves) accomplish a network of mathematical meanings taken-to-be-shared. From the observer's point of view, I call this network of meanings a mathematical "theme". In the episode presented, the theme is the relationship of materials to mathematical terms.

Because "a teacher is not safed against the students' creativity" (Bauersfeld, personal communication) the students can originally contribute to the theme so that the theme does not become a representation of the mathematical content which the teacher intended to establish. Realizing his/her intentions, the teacher is also dependent on the students' indications of understanding. Conversely, the students are dependent on the teacher's understanding and acceptance of their contributions. So, the theme is not a fixed body of knowledge, but as the topic of discourse, it is interactively constituted and changes through the negotiation of meaning. Over an extended period of time, the network of themes forms the realized curriculum (as opposed to the intended curriculum).

If the teacher does not direct and evaluate the students rigidly step-by-step and if the students originally contribute to the dialogue in the classroom, then the theme resembles a river that produces its own bed. The outcome of the dialogue is not clear from the outset. The following example serves as another illustration of how the students influence the theme which is contrary to the teacher's intention.

Before, the teacher wrote two equations on the blackboard: $4 = 3+1$ and $4 = 2+2$. In accordance with the equations, one had to fix stickers on the blackboard. Realizing these representations, the participants do not encounter immense conflicts. In the first example, 3 airplanes with a red tail and 1 airplane with a silvery tail were fixed on the board. In the second example, 2 mushrooms with a collar and 2 mushrooms without a collar were chosen.

The teacher then writes $4 = 1 + 3$ on the board, and immediately she fixes 5 apples: 1 apple with two worms printed on the apple, 3 apples each having one worm, and 1 apple without a worm – all worms are smiling cheerfully (Figure 2).

The teacher expects the students to realize that one apple, the right one, has to be removed. But the students begin a heated debate. The following sequence is only a small part of the discussion that occured:

Patrick:	There are four apples with one worm, and one does not have one.
Teacher:	There, kids have just put their hands up. Matthias!
Matthias:	There are two in the apple. And in the other three are only one. And there is one without. There, we have to write a zero because there is no worm inside. Zero worm.
Student D:	There, one has to write one.
Students:	(simultaneously shouting) No, one, no, one, no zero.
Student E:	A zero is if there is nothing at all.
Matthias:	Hey, hey, but there is no worm on it.

Unsuccessfully, the teacher tries to emphasize the counting of the apples. She gives suggestive hints, and she confirms students' statements that she expected. For example, the student Katrin, who often tries to assist the teacher, provides the teacher an opportunity to do such:

Katrin:	It is not about worms, it is about apples.
Teacher:	And what do you think if it is about apples?
Katrin:	Then, one has to write a four and there a one.
Matthias:	But there is not a single worm in it.

Finally, the participants determine that different mathematical statements fit the empirical phenomena. An alternative statement, "$5 = 2 + 1 + 1 + 1 + 0$," is written on the blackboard. Also, the participants agree that the first statement fits the empirical situation, if the right apple is taken away.

In this episode, the relationship between a pictorial representation and mathematics becomes explicitly ambiguous – diverging from the teacher's expectation and contrary to her initial attempts to direct all the students' attention to count the apples. A dispute begins about the validity of the mathematization. By the processes of negotiation, the participants recognize that mathematizations depend on specific foci of attention: Is one interested in apples or worms? Both perspectives are related to alternative mathematical statements.

4. REGULARITIES OF THE NEGOTIATION OF MEANING

Emergence of Intersubjectivity

In time, the negotiation of meaning forms commitments between the participants and stable expectations from the individual's point of view. In smooth interactions background knowledge is taken-to-be-shared. What was constituted explicitly before, now remains tacit. Cobb calls this process the institutionalization of knowledge (1990, pp. 211–213). Studying everyday life, ethnomethodologists point out that knowledge is confirmed to be shared and to be given by descriptive "accounting practices": "the stories that people are continually telling are descriptive accounts. ... To construct an account is to make an object or event (past or present) observable and understandable to oneself or to someone else. To make an object or event observable and understandable is to endow it with the status of an intersubjective object" (Leiter, 1980, pp. 161–162). Accordingly, in mathematics classrooms the participants' indications that something is the representation of a specific mathematical statement make it a representation of the mathematical statement.

Every empirical situation true-of-life (given as a story, picture, text, etc.) can be mathematized in various ways depending on one's interest. Therefore, an empirical situation is not a representation of a specific mathematical relationship, per se. The participants have to interpret the empirical situation accordingly so that it can be understood jointly as that representation. Through the process of negotiation, the teacher and the students endow the empirical situation with the status of mathematical coherence. In the episodes above, at first the empirical situations did not function as the intended representations, and intersubjectivity is jeopardized. This coherence is eventually established when the teacher and the students agree on several statements.

Context of Meaning

The ambiguity of single objects is reduced by relating their meaning to a context taken-to-be-shared. At the same time, the context is confirmed by constituting the meaning of the single object. The context and the singular meaning elaborate each other. From this ethnomethodological point of view, meanings are not given by the context of school mathematics that exists independently of the negotiation of meaning, but the context of school mathematics is continually constituted. Ethnomethodologists call this relationship "reflexivity" (Leiter, 1980, pp. 138–156). For example, first-graders experience that in the mathematics classroom apples, coloured blocks, chips, etc., are used differently than at home. The members of the classroom ascribe mathematical meanings to the things. At the same time, the meaning of what is called "math" becomes clearer to the first graders: One has to count chips, to assign apples to numerical symbols, and so on.

If the observer looks at the classroom life in the way an ethnographer does who investigates a strange culture, the observer might be astonished by what is taken for granted by the members of this classroom culture. However, in the treadmill of everyday life, the participants would say that they know what mathematics and the classroom practice really are. In everyday classroom situations, the teacher and the students often constitute the context routinely without conflicts and without being aware of the ongoing accomplishments. So, in the participants' experiences the context can seem to be pre-given. In everyday classroom practice the teacher and the students assume that the context is known which is, in fact, taken-to-be-shared, vague, and implicit.

In the above episode, the context of "addition" is obviously taken for granted. Nevertheless, it would be reasonable to interpret the apples as a representation of subtraction: $1 = 5 - 4$, only one apple is not wormy, or $3 = 4 - 1$, three wormy apples more than apples free of worms. Such differences of numbers have not been produced before in the classroom. Therefore, it is plausible that the participants do not realize them. The factual limitation of all realized possibilities is explained by the classroom practice and not only by the constraints of the empirical facts.

In contrast to the observer's point of view, the participants may not have to experience the discourse processes as limiting their creativity. If in a usual classroom a participant presented many of these alternatives, s(he) might provoke irritations, might be accused of straying from the theme, or might be valued as being unfamiliar with school mathematics. In the classroom observed, such cases of irritation are seldom. Presumably, the students try to figure out the teacher's expectation, and the teacher can be confident that the students develop a feeling for the context taken for granted by her.

Observing such classrooms, the educator might want that the participants were interested in constituting many different mathematizations. The context taken for granted by the participants could be wider, but the discourse can not be free of context.

Routines

Several microethnographical studies explore implicit regularities of classroom interaction. Studies of the stability of regularities have been motivated by the disillusionment with regard to the period of educational reform in the 1960s and 1970s (cf. Maier and Voigt, 1992):

"Routines" function towards minimizing the permanent risk of a possible collapse and disorganization of the classroom discourse. In several case studies, Voigt (1989) has reconstructed teachers' and students' routines in which a smooth functioning of the classroom discourse proceeds and mathematical meanings are interactively constituted.

Obligations

The routines are connected by interactional "obligations." A routine action lets the observer expect another routine action in the form of a reaction if no conflict between the participants arises. The obligations also become obvious in cases of conflict. In the following episode of another classroom, a student does not fulfill the teacher's expectation; he violates an obligation from the observer's point of view. The teacher takes care to maintain the sense of normality and the image of the classroom oriented towards the folk ideal of discovery learning:

> Teacher: That is enough for the moment. We cannot write down all the results, don't you think? Does anybody notice anything?
>
> Student: What am I supposed to notice?
>
> Teacher: What are you supposed to notice? That's something you ought to know yourself. Björn, have you noticed anything?

In addition, the teacher's activities are constrained by obligations. For example, in traditional classrooms the students often expect the teacher to present an official algorithm of solving problems step-by-step without the need for reflection (what to do next?). So, the students are not only the "victims" of this classroom culture but also are the "culprits."

Patterns of Interaction

The network of routines and obligations can be described as "patterns of inter-action" (cf. Bauersfeld, 1988; Jungwirth, 1991; McNeal, 1991; Voigt, 1985). The patterns of interactions are considered as regularities interactively constituted and not as fixed rules: "What is presented is a level on which processes remain processes and do not coagulate into entities, to which the very process from which they are abstracted is assigned to as effect" (Falk and Steinert, 1973, p. 20).

For example, Voigt (1985) reconstructs the "elicitation pattern" which hints to the contradictory combination of two claims. The idea of eliciting a clear-cut body of mathematical knowledge is associated with the claims of a liberal and child-centred classroom. In the elicitation pattern three phases can be distinguished:
- The teacher sets up an ambiguous task, the students offer different answers which the teacher evaluates preliminarily.
- If the students' contributions are too divergent the teacher guides the students towards one definite argument, solution, etc. Believing to help the students, the teacher asks small questions to elicit bits of knowledge.
- The teacher and the students reflect and evaluate what has been done.

The stability of the elicitation pattern can be explained partly by the participants' routines. Jungwirth (1991) reconstructs gender-specific routines by which boys and girls contribute to the elicitation pattern and modify the ordinary elicitation pattern, respectively. Because the boys participate more successfully in the smooth accomplishment of the first two phases of the pattern they might erroneously appear to the teachers to be more mathematically competent.

Stability of the Pattern of Interaction

When educators study classroom processes using microethnographical means, the stability of traditional regularities is astonishing. While one might expect these regularities to be structured by rational argumentations, the mathematical discourses are actually highly socially structured. Even in classroom situations which are expected to be designed to represent modern educational claims, traditional patterns are reconstructed. For example, students' group work is implemented in order to overcome the bad features of frontal teaching. Bauersfeld (1982) uses the concept "habitus" in order to explain the persistence of students' routines during an episode of group work. He demonstrates that in this group work the students constitute patterns and ways of solving the mathematical problem which partly appear as copies of traditional frontal teaching. Changing the formal social organization of classroom life alone does not guarantee the immediate change of hidden and stable regularities in the microprocesses. Among others, social norms of argumentation have to be changed during frontal teaching in order to stimulate the students to change their habitus during group work (cf. Wood and Yackel,

1990). The resistence to the changing of the regularities has to be taken into account when changing the micro-culture of mathematics classrooms.

Voigt (1989) points out that teachers neither realize that traditional patterns of interaction are still alive in their classrooms nor that they contradict the teacher's intentions. The tradition of "Socratic catechism" still has an effect in the microculture today (cf. Maier and Voigt, 1989). In the context of pre-service teacher training, examined "under the microscope," ideal teaching styles, in part, merely seem to be staged in "holiday lessons" (cf. Andelfinger and Voigt, 1986). Presumably, in everyday classroom processes teachers reproduce routines and background understandings which have been unintentionally developed during their schooldays.

Therefore, Yackel et al. (1990) influence classroom life, for example, by encouraging teachers to change specific social norms in the classroom discourse. Voigt (1991) develops teacher training courses where teachers videotape their own teaching and analyze the tapes by themselves.

Nevertheless, it is to be taken into account that in everyday classroom life routines and patterns cannot be eliminated at all. Because of the permanent ambiguity we want assurance, relief, implicit orientation, and reliability. "We like to settle down like in a familiar nest, the nest of everyday life" (duBois-Reymond and Söll, 1974, p 13).

Pattern of Direct Mathematization

Previously, an illustration was presented which shows an ape, a keeper, and bananas. Using the picture, the participants in the observed elementary school classroom derived a specific, step-by-step solution. When the teacher became irritated by diverse suggestions (one more banana, you have to add, etc.) the teacher often asked a sequence of questions:

> How many bananas has the keeper brought? Answer: 5.
> Which sign has to be written if something is taken away?
> Answer: Minus.
> How many bananas does the keeper give to the ape? Answer: 2.
> The equals sign is already there.
> How many bananas has the keeper left over? Answer: 3.

Through this procedure the picture becomes a specific arithmetical task, and a "number sentence" represents the solution. Details of the picture becomes clearly related to mathematical symbols (signs). The sequence of questions occurs concurrently with writing the sentence. Accordingly, nearly all pictures in the textbook are stereotyped as such: the intended subtraction is indicated by the picture of persons or things leaving the picture from the right side. In subsequent lessons, the students learn so that, in the long run, they can independently mathematize similar pictures accordingly.

Producing the mathematical statements expected by the teacher, the students do not necessarily interpret the picture in terms of numerical part-whole relations.

It is possible that students learn how to relate parts of stereotyped pictures to single numerals step by step without thinking about the relationship among numerals.

This procedure is an example of a pattern of interaction which will be termed the pattern of "direct mathematization." The pattern of direct mathematization is a "thematic pattern" because it is specific to mathematics classrooms and describes the development of a theme. Producing a thematic pattern, the teacher and the students constitute the theme routinely. Alternative interpretations of the empirical situations do not become thematic usually. Episodes, in which alternatives become valid such as the episodes in the previous sections, are exceptions of the classroom observed. The pattern of direct mathematization can be reconstructed in most of the classroom episodes in which empirical situations are interpreted mathematically and in which the students are not familiar with these types of problems.

After the lesson in which the apples and worms have been counted, the teacher discussed the lesson with the observer. The teacher was unhappy, and she suspected that the scene had gone wrong because the task was ambiguous. The observer objected that every empirical situation could be mathematized differently and that a one-to-one relationship of mathematical symbols and empirical phenomena would not be intellectually honest. Nevertheless, during the discussion between the observer and the teacher, the teacher initially did not interpret the scene as an intellectual challenge.

In sum, in the classroom situation, objects are ambiguous. Because of the differences between their background knowledge, the teacher and the students must negotiate mathematical meanings. During negotiation the single meaning and the context of meaning inform each other. The potential conflicts during negotiation are minimized through routines and obligations. The relations among routines and obligations form (thematic) patterns of interaction. Through these patterns, the teacher and the students arrive at mathematical meanings taken-to-be-shared without realizing all alternative interpretations of the objects. In everyday classroom life, there is the danger that the processes degenerate into proceduralized rules and rituals. However, at the microlevel, the regularities are not pregiven but the negotiation could be improvised at every moment. The regularities are the accomplishment of the participants during their interactions. Further, stereotyped textbooks, tests, the persistence of tradition, and the need for reliability might contribute to their stability.

5. INDIRECT RELATIONS BETWEEN SOCIAL INTERACTION AND MATHEMATICS LEARNING

The relationship between social interaction and learning mathematics can be studied from several perspectives. In the following discussion, an attempt will be made to compare two basic theoretical approaches roughly. Then, an interactionist approach will be proposed which attempts to mediate between these two basic approaches. Finally, from this interactionist perspective, more concrete conclusions will be drawn with regard to the topics of the preceding sections.

The Focus on the Individual

For a long period of time, research in mathematics education has been pro-
foundly influenced by Piaget's genetic epistemology and developmental psychology. Many researchers focus their attention on the cognitive development of
individual children (cf. Steffe et al., 1983). The child's mathematical knowledge
is viewed as the product of the individual's conceptual operations. Currently, this
perspective is well known as radical constructivism (cf. von Glasersfeld, 1987).
From this point of view the individual's knowledge can be at best viable. That
is, if the student's knowledge is compatible with new experiences subjectively
interpreted, the student's knowledge is confirmed without necessarily being true
or intended by the teacher. In this perspective there is no place for the idea
of transmission of knowledge from the teacher to the student or of the idea of
interpreting the reality as it is.

Although the sociological aspects in Piaget's research are de-emphasized Piaget remarks that "social interaction is a necessary condition for the development
of logic" (cited by Doise and Mugny, 1984, p. 19). In Geneve, Piaget's followers,
Doise and Mugny (1984) and Perret-Clermont (1980), use Piaget's developmental
psychology as a basis of their work. They explore social interactions between
children, and study conflicts between the children's perspectives as conditions for
mental reorganizations.

Perret-Clermont (1980) views learning as the result of the learner's attempt
to resolve conflicting points of view. At first, the conflicting points of view
are indicated as contradictions between arguments of different participants in the
interaction. This socioconflict can result in a contradiction of one's own point
of view which leads to mental reorganization. Kumagai (1988) stresses that the
different perspectives held up by the individuals as they interact form a helpful
prerequisite for learning mathematics. That the observer realizes an interactional
conflict does not ensure the occurence of learning. Balacheff (1986) suggests
one must consider whether the learner recognizes that a contradiction exists.
The ability to recognize a contradiction depends on the learner's background
knowledge.

Solomon (1989, pp. 109–119) criticizes the extension of Piaget's work by
his followers because they view only the acquisition of knowledge and not the
knowledge itself as socially conditioned. If we do not assume innate organizers,
how can we think of intersubjective meanings when each individual constructs
his/her own meaning? The cultural aspects of mathematical practices in school
have to be taken into account. "Acculturation and the institutionalization of
mathematical practices are ... a necessary aspect of children's mathematical
education. Analyses that focus solely on individual children's construction of
mathematical knowledge tell only half of a good story" (Cobb, 1990, p. 213; see
also Bauersfeld, 1980).

Bruner is well-known for emphasizing children's discoveries and for criticizing
the teacher's guidance. In 1986, he remarks on a development in his own work:
"I have come increasingly to recognize that most learning in most settings is a

communal activity, a sharing of the culture" (p. 127). Twenty years earlier, Bruner (1966) wrote: "Culture, thus, is not discovered; it is passed on or forgotten. All this suggests to me that we have better to be cautious in talking about the method of discovery, or discovery as the principal vehicle of education" (p. 101). Edwards and Mercer (1987) increase the importance of culture as a point of reference in educational research: "The child-centred ideology needs to be replaced with one that emphasizes the socio-cultural and discursive bases of knowledge and learning" (p. 168). Instead of the individual person, now, the social group takes the place as the basic element of interest. Before discussing the appropriateness of this replacement, the implications of a one-sided emphasis on culture will be roughly outlined.

In contrast to locating rational argumentations in the private realms of individual experience, Harré (1984) understands rationality as "a feature of public-collective discourses to which there may have been several individual contributions" (p. 120). "To say that someone is rational is not to congratulate them on their private cognitive processes but to praise them for their contributions to the collective discourse" (p. 119). Like Wittgenstein (1967), Harré and other social constructivists look for the roots of (mathematical) competence in social activities. Similarly, Solomon (1989) claims that the cognitive development is "the progressive socialization of the child's judgements" (p. 118). In mathematics lessons "learning is the initiation into a social tradition" (p. 150).

Several researchers, inspired by Vygotsky's work (1978), study social events in order to explore links between culture and cognition. Vygotsky assumes that the characteristics of adult-guided interactions are internalized by the learner in development: "all higher mental functions are internalized social relationships" (1981, pp. 163–164). Vygotsky characterizes the learner's development in terms of shifts in control and responsibility. The more competent participants guide the interactions so that learners can participate in activities that they could not otherwise master by themselves. More and more, during these interactions, learners increase their control and responsibility.

The Focus on Interaction – a Chance of Mediating the Foci on the Subject and on Culture

Upon to this point in the discussion, several theoretical approaches with regard to learning mathematics through interaction have been mentioned. Crudely compared, two antithetical strands can be reconstructed: individualism versus collectivism. With reference to Piaget (especially in von Glasersfeld's interpretation), learning mathematics is viewed as structured by the individual's attempts to resolve what the individual finds problematic in the world of his/her experience. With reference to Vygotsky (especially in Leont'ev's interpretation) the given environment seems to direct the individual's learning of mathematics. On the one hand, the individual is the actor (subject), and the mathematical knowledge is constructed by the actor. On the other hand, the individual is the object of cultural practices, and given mathematical knowledge is internalized.

Of course, the comparison gives a crude contrast of opposing tendencies. Each school has produced sophisticated ideas to answer questions posed by the other. Bauersfeld (1988, pp. 38–40) and Cobb (1990, pp. 213–215) stress the complementary character of these theories.

The author considers the emphasis on the subject's experience as the starting-point in order to understand the relationship of interaction and learning. The primary normative reason is that concepts like "socialization," "internalization," "initiation into a social tradition," etc., do not sufficiently explain what the author considers to be the most important objective of mathematics education: "Bildung". Bildung is a main claim in the German tradition of thinking about education. Immanuel Kant (1783), the German philosopher of Enlightenment, criticizes what the individual considers to be habitual or natural. The individual should act on rational grounds in the individual's mind without relying on the other's guidance. (Today we would comment that reason, "Vernunft", is not innate but emerges when the subject becomes a member of culture.) The prominent objective of mathematics education is not that the students produce objectively solutions to mathematical problems, rather, that they do so insightfully and by using reasonable thinking. What on the behavioral level does not, in fact, make a difference should be an important subjective difference. Do the students act as expected because they intend to fulfill the teacher's expectations in order to participate successfully or because they draw conclusions in order to solve a mathematical problem in their experiental world?

Difference Instead of Deficit of Knowledge

From the interactionist point of view, the negotiation between the teacher and the students is a fascinating unit of analysis because of the participants' different perspectives. The teacher represents mathematical claims as well as the tradition of mathematics education whereas the student has a different background knowledge. The negotiation of meaning is a necessary condition for learning if the students' background knowledge differs from the knowledge the teacher wants the students to gain. This difference (not necessarily deficit) characterizes discourse in the classroom. This is particularly evident when the students erroneously assume to understand an ambiguous topic mathematically according to the teacher's intentions. Describing this situation, Griffin and Mehan (1981) uses a trenchant phrase: "Thus, instead of making entries on a blank slate, teaching in school seems to be involved in erasing entries from a too full slate" (p. 212).

Nevertheless the student is not a minor partner at all. In the classroom episodes presented previously, the students do not fulfill the teacher's previous expectations, but, finally, they can convince the teacher. The students' contributions to the mathematical theme have to be considered as an essential aspect in the theory: the students' thinking and the mathematical theme develop reflexively; the student's learning contributes to the evolution of the theme which contributes to the student's learning.

In the German classroom observed, the scene with the apples and worms had

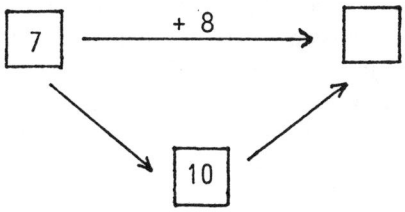

Fig. 3.

a long-term effect on the teacher. More often than before, the teacher began to realize that many of her tasks were ambiguous and that often the students' strange solutions should not be evaluated as wrong or as diverging from the task, but only diverging from the teacher's own limited interpretation of the task. After such classroom situations, the teacher came to the observer in the back of the classroom, and remarked that she had posed an "apple-and-worm-task" once more. Also, the teacher began to take an interest in mathematical problems which explicitly required students to look for alternative interpretations of pictorial representations (cf. Wittmann and Müller, 1990).

Pattern of Interaction and Learning

One the one hand, the student's participation in patterns of interaction can contribute to the student's cognitive development. Referring to Bruner (1976), Krummheuer (1991) analyzes mathematics classrooms by reconstructing specific patterns of interaction, "formats of argumentation," which are initiated by the teacher and which become obliging for the students. The more often the students participate successfully in these formats the more often the teacher will let the students take on responsibility. Eventually, the students gain the ability to argue independently.

On the other hand, the conclusion should not be drawn that the teacher, in principle, must force the students to participate in specific patterns. The students can by-pass the regulations enforced by the teacher:

In German classrooms a specific diagram is used as an aid to train students in carrying the ten: $7 + 8 =$:

8 is partitioned as 3 (in order to fill 7 up to 10) and 5. The students are forced to complete the diagram, that is, mark the left array with "+3" and the right array with "+5" and note the solution.

Instead, many students solve the arithmetical task by their own methods mentally to complete the right square. Finally, they mark the arrays because they know that the teacher wants them to do so.

Mead (1934), a founder of symbolic interactionism, describes the subject's identity as a dynamic balance between the "I" and the "Me." The student has to keep a balance between what the student wants to do, what the student can do, what the student perceives s/he is expected to do, and what the student perceives others do. Similar in the last example, there may be a difference between the student's private sense making and the student's reasons for behavior. Therefore, the emphasis on the negotiation of mathematical meaning seems to be more promising than the tradition of ritualized classroom interaction controlled by the teacher, step-by-step.

The thematic pattern of direct mathematization can prompt the students to produce even wrong mathematical statements although each step may seem reasonable. For example, with regard to the picture of the ape, the teacher could replace the last question with "How many more bananas has the keeper than the ape?" Answering all the teacher's questions correctly, the student contributes to a mathematically incorrect statement: $5 - 2 = 1$. Thus, the smooth participation in a pattern of interaction does not necessarily contribute to critical thinking.

Up to this point, interactional regularities as well as the subject's creativity have been described as prerequisites for learning mathematics. "So, what we have here are neither automatic rituals – repeated endlessly and mechanically, nor instantaneous creations, – emerging uniquely upon each occasion of interaction. These are negotiated conventions – spontaneous improvisations on basic patterns of interaction" (Griffin and Mehan, 1981, p. 205).

Conventionalized Relations and Learning

Conventions can be useful as well as the cause of obstacles in learning. On the one hand, conventions enable the participants to also collaborate even if the individuals' background knowledge differ. In analyzing mathematical discussions, Walther (1982) and Lampert (1990) take conventions as the means and products of the negotiation of mathematical meaning. Both authors describe how teachers take care of conventions which are constituted on the basis of the students' contributions in the class discussion. Here the teacher acts as a mediator between students' individual knowledge and school mathematics.

On the other hand, conventions, which are taken for granted, could become a burden to the learner: In the German classroom observed, during the beginning of the schoolyear the first graders have experienced that numbers always stand for the amounts of several things or persons. Some months later, the teacher presents several columns filled with different stickers, each column consists of uniform stickers. At the top of each column the price of one sticker of the column is written (e.g., 4 c.). In order to construct addition problems, the teacher requires the students to purchase several stickers. But the students protest vehemently. They demand that the numbers at the top of the columns are changed. They argue that every written number has to represent the quantity of stickers of the corresponding column. Although the teacher explains her intention a student extorts a long and excited negotiation of meaning. Another well-known effect of the direct relation between numbers and empirical phenomena usually becomes problematic in the introduction of negative numbers (cf. Brousseau, 1983; Schubring, 1988). The learner can experience the statement that -4 is "more" than -5 as inconsistent with the argument that 4 apples are less than 5. Also, addition of negative numbers does not signify "increase" although, in the elementary school, adding is made plausible as increasing an amount.

Students must learn to distinguish between mathematical arguments and those conventions of mathematizing which are established in order to make the negotiation of mathematical meaning easier in school. This is a difficult point, because mathematics itself is full of conventions invented to make the mathematical communication easier. Nevertheless in school mathematics, there are many conventions that educators have developed for didactical reasons and are short lived during teaching mathematics. Insofar as the student's sceptical attitude towards the routine participation in regularities is reasonable, the student should share the responsibility to negotiate mathematical meanings of the objects of classroom discourse.

In the pattern of direct mathematization mathematical symbols (signs) become closely related to empirical phenomena. Perhaps, by participating in this pattern the so-called weaker student might gain a sense of understanding of the mathematical symbols. However, if we suspect that the weaker student's problem is not the inability to reason mathematically but interpretations (mathematizations) diverging from the conventions taken for granted by the teacher, the pattern of direct mathematization does not support the student's mathematical progress. The negotiation of meaning is helpful when used to improvise the relationship between empirical situations and mathematical statements in a flexible manner.

Education for the Negotiation of Mathematical Meanings

Finally, this discussion raises an additional issue. In the mathematics classroom students do not only learn mathematics, they also learn to negotiate mathematical meanings with an expert, the teacher. Presumably, the student-teacher-interaction is analoguous to the relationship between the layman and the expert paradigmatically, and anticipates it. Does the student experience the teacher as a dogmatic authority who determines mathematizations of empirical situations? Or, does the student collaborate with a teacher who recognizes and endures the tension between the layman's and the expert's different mathematizations?

As a future citizen, the student will have to cope with many problems which are mathematically not transparent or which s/he would spontaneously mathematize differently from the expert (e.g., life insurance, limits of poison in the environment, taxes). Is the citizen competent in this situation to distinguish between the necessary mathematical inferences and the assumptions of modelling which depend on interests? There is the hope that if we take care of the quality of the negotiation of mathematical meaning we could improve the culture of the mathematics classroom as well as the education of the "competent layman."

NOTE

Several aspects discussed in the paper were elaborated in the course of discussions with Heinrich Bauersfeld, Götz Krummheuer, and Angelika Neth (Germany) and with Paul Cobb, Terry Wood, Erna Yackel (U.S.A.). I especially thank John Richards, Analúcia Dias Schliemann, and Erna Yackel for their written reactions to an earlier version presented at ICME-7.

The research reported in this paper was supported by the Spencer Foundation. The opinions expressed do not necessarily reflect the views of the foundation.

REFERENCES

Andelfinger, B. and Voigt, J.: 1986, 'Vorführstunden und alltäglicher Mathematikunterricht – Zur Ausbildung von Referendaren im Fach Mathematik (S I/S II)', *Zentralblatt für Didaktik der Mathematik* 1, 2–9.

Balacheff, N.: 1986, 'Cognitive versus situational analysis of problem-solving behaviors', *For the Learning of Mathematics* 6(3), 10–12.

Balacheff, N. and Laborde, C.: 1988, 'Social interactions for experimental studies of pupils' conceptions: Its relevance for research in didactics of mathematics', in: H.-G. Steiner and A. Vermandel (eds.), *Foundations and Methodology of the Discipline Mathematics Education*. Proceedings of the 2nd TME-Conference, IDM, Bielefeld, pp. 189–195.

Bartolini-Bussi, M.: 1990, 'Mathematics knowledge as a collective enterprise', in: *Proceedings of the 4th SCTP Conference* in Brakel, West Germany, IDM, Bielefeld, 121–151.

Bauersfeld, H.: 1980, 'Hidden dimensions in the so-called reality of a mathematics classroom', *Educational Studies in Mathematics* 11, 23–41.

Bauersfeld, H.: 1982, 'Analysen zur Kommunikation im Mathematikunterricht', in H. Bauersfeld et al. (ed.), *Analysen zum Unterrichtshandeln*, Aulis, Köln, pp. 1–40.

Bauersfeld, H.: 1988, 'Interaction, construction, and knowledge – Alternative perspectives for mathematics education', in T. Cooney and D. Grouws (eds.), *Effective Mathematics Teaching*, N.C.T.M., Reston/Virginia, pp. 27–46.

Bauersfeld, H. and Krummheuer, G., and Voigt, J.: 1988, 'Interactional theory of learning and teaching mathematics and related microethnographical studies', in H.-G. Steiner and A. Vermandel (eds.), *Foundations and Methodology of the Discipline Mathematics Education*, Antwerp, 174–188.

Bishop, A. J. and Geoffree, F.: 1986, 'Classroom organisation and dynamics', in B. Christiansen, A. G. Howsen, and M. Otte (eds.), *Perspectives on Mathematics Education*, Reidel, Dordrecht, 309–365.

Blumer, H.: 1969, *Symbolic Interactionism: Perspective and Method*, Prentice-Hall, Englewood Cliffs.

Brousseau, G.: 1983, 'Les obstacles épistémologiques et les problèmes en mathématiques', *Recherches en didactique des mathématiques* 2, 164–197.

Bruner, J.: 1966, 'Some elements of discovery', in L. S. Shulman and E. R. Keisler (eds.), *Learning by Discovery: A Critical Appraisal*, Rand McNally, Chicago, 101–113.

Bruner, J.: 1976, 'Early rule structure: The case of peek-a-boo', in R. Harré (ed.), *Life Sentences*, Wiley and Sons, London.

Bruner, J.: 1986, *Actual Minds, Possible Worlds*, Harvard University Press, Cambridge/Mass.

Carraher, T. N., Carraher, D. W., and Schliemann, A. D.: 1985, 'Mathematics in the streets and in schools', *British Journal of Developmental Psychology* 3, 21–29.

Cobb, P.: 1990, 'Multiple perspectives', in L. P. Steffe and T. Wood (eds.), *Transforming Children's Mathematical Education: International Perspectives*, Lawrence Erlbaum Association, Hillsdale, N.J., 200–215.

Cobb, P., Wood, T., and Yackel, E.: 1991, 'A constructivist approach to second grade mathematics', in E. von Glasersfeld (ed.), *Constructivism in Mathematics Education*, Kluwer, Dordrecht, 157–176.

Doise, W. and Mugny, G.: 1984, *The Social Development of the Intellect*, Pergamon, Oxford.

duBois-Reymond, M. and Söll, B.: 1974, *Neuköllner Schulbuch, 2. Band*, Suhrkamp, Frankfurt/M.

Edwards, D. and Mercer, N.: 1987, *Common Knowledge. The Development of Understanding in the Classroom*, Methuen, London.

Falk, G. and Steinert, H.: 1973, 'Über den Soziologen als Konstrukteur von Wirklichkeit, das Wesen der sozialen Realität, die Definition sozialer Situationen und die Strategien ihrer Bewältigung', in H. Steinert (ed.), *Symbolische Interaktion. Arbeiten zu einer reflexiven Soziologie*, Klett, Stuttgart, pp. 13–46.

Garfinkel, H.: 1967, *Studies in Ethnomethodology*, Prentice-Hill, New Jersey.

Goffman, E.: 1974, *Frame Analysis – An Essay on the Organization of Experience*, Harvard University Press, Cambridge.

Griffin, P. and Mehan, H.: 1981, 'Sense and ritual in classroom discourse', in F. Coulmas (ed.), *Conversational Routine*, Mounton, The Hague, pp. 187–214.

Harré, R.: 1984, *Personal Beeing. A Theory for Individual Psychology*, Harvard University Press, Cambridge.

Jungwirth, H.: 1991, 'Interaction and gender – findings of a microethnographical approach to classroom discourse', *Educational Studies in Mathematics* 22, 263–284.

Krummheuer, G.: 1983, *Algebraische Termumformungen in der Sekundarstufe I – Abschlußbericht eines Forschungsprojektes*, Materialien und Studien Bd. 31, IDM, Bielefeld.

Krummheuer, G.: 1991, 'Argumentationsformate im Mathematikunterricht', in H. Maier and J. Voigt (eds.), *Interpretative Unterrichtsforschung*, Aulis, Köln, 57–78.

Kumagai, K.: 1988, 'Research on the 'sharing process' in the mathematics classroom – An attempt to construct a mathematics classroom', *Tsukaba Journal of Educational Study in Mathematics Japan* 7, 247–257.

Lakatos, J.: 1976, *Proofs and Refutations – The Logic of Mathematical Discovery*, Cambridge University Press, London.

Lampert, M.: 1990, 'Connecting inventions with conventions', in L. P. Steffe and T. Wood (eds.), *Transforming Children's Mathematics Education*, Lawrence Erlbaum, Hillsdale/N.J., 253–265.

Leiter, K.: 1980, *A Primer on Etnomethodology*, Oxford University Press, New York.

Maier, H. and Voigt, J.: 1989, 'Die entwickelnde Lehrerfrage im Mathematikunterricht, Teil I', *Mathematica Didactica* 12, 23–55.

Maier, H. and Voigt, J.: 1992, 'Teaching styles in mathematics education', *Zentralblatt für Didaktik der Mathematik* 7, 249–253.

McNeal, M. G.: 1991, *The Social Context of Mathematical Development*, Doctoral dissertation, Purdue University.

Mead, G. H.: 1934, *Mind, Self and Society*, University of Chicago Press, Chicago.

Mehan, H.: 1979, *Learning Lessons*, Harvard University, Cambridge.

Neth, A. and Voigt, J.: 1991, 'Lebensweltliche Inszenierungen. Die Aushandlung schulmathematischer Bedeutungen an Sachaufgaben', in H. Maier and J. Voigt (eds.), *Interpretative Unterrichtsforschung*, Aulis, Köln, 79–116.

Perret-Clermont, A.-N.: 1980, *Social Interaction and Cognitive Development in Children*, Academic Press, London.

Saxe, G. B.: 1990, *Culture and Cognitive Development: Studies in Mathematical Understanding*, Lawrence Erlbaum, Hillsdale, N.J.

Schubring, G.: 1988, 'Historische Begriffsentwicklung und Lernprozeß aus der Sicht neuerer mathematikdidaktischer Konzeptionen (Fehler, obstacles, Transposition)', *Zentralblatt für Didaktik der Mathematik* 20, 138–148.

Solomon, Y.: 1989, *The Practice of Mathematics*, Routledge, London.

Steffe, L. P., von Glasersfeld, E., Richards, J., and Cobb, P.: 1983, *Children's Counting Types: Philosophy, Theory, and Application*, Praeger Scientific, New York.

Steinbring, H.: 1991, 'Mathematics in teaching processes. The disparity between teacher and student knowledge', *Recherches en didactique des mathématiques* 11(1), 65–108.

Steinbring, H.: 1993, 'The relation between social and conceptual conventions in everyday mathematics teaching', in L. Bazzini and H.-G. Steiner (eds.), *Proceedings of the Second Italian–German Bilateral Symposium on Didactics of Mathematics* (in press).

Struve, H.: 1990, 'Analysis of didactical developments on the basis of rational reconstructions', *Proceedings of the BISME-2*, Bratislava, 99–119.

Struve, R. and Voigt, J.: 1988, 'Die Unterrichtsszene im Menon-Dialog – Analyse und Kritik auf dem Hintergrund von Interaktionsanalysen des heutigen Mathematikunterrichts', *Journal für Mathematik-Didaktik* 9(4), 259–285.

Voigt, J.: 1985, 'Pattern und routines in classroom interaction', *Recherches en Didactique des Mathématiques* 6(1), 69–118.

Voigt, J.: 1989, 'Social functions of routines and consequences for subject matter learning', *International Journal of Educational Research* 13(6), 647–656.

Voigt, J.: 1990, 'The microethnographical investigation of the interactive constitution of mathematical meaning', *Proceedings of the 2nd Bratislava International Symposium on Mathematics Education*, 120–143.

Voigt, J.: 1991, 'Interaktionsanalysen in der Lehrerfortbildung', *Zentralblatt für Didaktik der Mathematik* 5, 161–168.

von Glasersfeld, E.: 1987, *Wissen, Sprache und Wirklichkeit. Arbeiten zum radikalen Konstruktivismus*, Vieweg, Braunschweig.

Vygotsky, L. S.: 1978, *Mind in Society*, Harvard University, Cambridge, MA.

Vygotsky, L. S.: 1981, 'The genesis of higher mental functions', in J. V. Wertsch (ed.), *The Concept of Activity in Soviet Psychology*, Sharpe, Armonk.

Walkerdine, V.: 1988, *The Mastery of Reason*, Routledge, London.

Walther, G.: 1982, 'Acquiring mathematical knowledge', *Mathematics Teaching* 101, 10–12.

Wittgenstein, L.: 1967, *Remarks on the Foundations of Mathematics*, Blackwell, Oxford.

Wittmann, E. Ch. and Müller, G. N.: 1990, *Handbuch produktiver Rechenübungen, Band 1.*, Klett, Stuttgart.

Wood, T. and Yackel, E.: 1990, 'The development of collaborative dialogue in small group interactions', in L. P. Steffe and T. Wood (eds.), *Transforming Children's Mathematics Education*, Lawrence Erlbaum Association, Hillsdale, N.J., 200–215.

Yackel, E., Cobb, P., Wood, T., Wheatley, G., and Merkel, G.: 1990, 'The importance of social interaction in children's construction of mathematical knowledge', in T. J. Cooney (ed.), *Teaching and Learning Mathematics in the 1990's*, NCTM, Reston, pp. 12–21.

Universität Hamburg,
Fachbereich Erziehungswissenschaft,
Von-Melle-Park 8,
20146 Hamburg 13,
Germany